AF358738

Empowering Materials Processing and Performance from Data and AI

Empowering Materials Processing and Performance from Data and AI

Editors

Francisco Chinesta
Elías Cueto
Benjamin Klusemann

MDPI • Basel • Beijing • Wuhan • Barcelona • Belgrade • Manchester • Tokyo • Cluj • Tianjin

Editors

Francisco Chinesta
PIMM Laboratory
ESI Group Chair
Arts et Métiers Institute
of Technology
Paris
France

Elías Cueto
Aragon Institute of
Engineering Research
Universidad de Zaragoza
Zaragoza
Spain

Benjamin Klusemann
Institute of Product and
Process Innovation
Leuphana University of Lüneburg
Lüneburg
Germany

Editorial Office
MDPI
St. Alban-Anlage 66
4052 Basel, Switzerland

This is a reprint of articles from the Special Issue published online in the open access journal *Materials* (ISSN 1996-1944) (available at: www.mdpi.com/journal/materials/special_issues/empower_mat_process_perform_data_AI).

For citation purposes, cite each article independently as indicated on the article page online and as indicated below:

LastName, A.A.; LastName, B.B.; LastName, C.C. Article Title. *Journal Name* **Year**, *Volume Number*, Page Range.

ISBN 978-3-0365-1899-2 (Hbk)
ISBN 978-3-0365-1898-5 (PDF)

Contents

About the Editors . . . . **vii**

Francisco Chinesta, Elias Cueto and Benjamin Klusemann
Empowering Materials Processing and Performance from Data and AI
Reprinted from: *Materials* **2021**, *14*, 4409, doi:10.3390/ma14164409 . . . . **1**

Xinyi Gong, Yuksel C. Yabansu, Peter C. Collins and Surya R. Kalidindi
Evaluation of Ti–Mn Alloys for Additive Manufacturing Using High-Throughput Experimental
Assays and Gaussian Process Regression
Reprinted from: *Materials* **2020**, *13*, 4641, doi:10.3390/ma13204641 . . . . **5**

Norbert Huber
A Strategy for Dimensionality Reduction and Data Analysis Applied to
Microstructure–Property Relationships of Nanoporous Metals
Reprinted from: *Materials* **2021**, *14*, 1822, doi:10.3390/ma14081822 . . . . **25**

Minyoung Yun, Clara Argerich, Elias Cueto, Jean Louis Duval and Francisco Chinesta
Nonlinear Regression Operating on Microstructures Described from Topological Data Analysis
for the Real-Time Prediction of Effective Properties
Reprinted from: *Materials* **2020**, *13*, 2335, doi:10.3390/ma13102335 . . . . **49**

Juan Luis de Pablos, Edoardo Menga and Ignacio Romero
A Methodology for the Statistical Calibration of Complex Constitutive Material Models:
Application to Temperature-Dependent Elasto-Visco-Plastic Materials
Reprinted from: *Materials* **2020**, *13*, 4402, doi:10.3390/ma13194402 . . . . **61**

Alexander Hartmaier
Data-Oriented Constitutive Modeling of Plasticity in Metals
Reprinted from: *Materials* **2020**, *13*, 1600, doi:10.3390/ma13071600 . . . . **83**

Xiaoxin Lu, Julien Yvonnet, Leonidas Papadopoulos, Ioannis Kalogeris and Vissarion Papadopoulos
A Stochastic FE^2 Data-Driven Method for Nonlinear Multiscale Modeling
Reprinted from: *Materials* **2021**, *14*, 2875, doi:10.3390/ma14112875 . . . . **101**

David González, Alberto García-González, Francisco Chinesta and Elías Cueto
A Data-Driven Learning Method for Constitutive Modeling: Application to Vascular
Hyperelastic Soft Tissues
Reprinted from: *Materials* **2020**, *13*, 2319, doi:10.3390/ma13102319 . . . . **125**

Frederic E. Bock, Sören Keller, Norbert Huber and Benjamin Klusemann
Hybrid Modelling by Machine Learning Corrections of Analytical Model Predictions towards
High-Fidelity Simulation Solutions
Reprinted from: *Materials* **2021**, *14*, 1883, doi:10.3390/ma14081883 . . . . **143**

About the Editors

Francisco Chinesta

Francisco Chinesta is currently a full professor of computational physics at ENSAM (Paris, France), an honorary fellow of the Institut Universitaire de France and a fellow of the Spanish Royal Academy of Engineering. He is the president of the ESI Group scientific committee. He was an AIRBUS chair professor, and since 2013, he has been an ESI Group chair professor on advanced modeling and simulation of materials, structures, processes and systems. He has received many scientific awards: IACM Zienkiewicz and the ESAFORM in four different fields: bio-engineering, material forming processes, rheology and computational mechanics (model order reduction and engineered artificial intelligence). He is the author of about 320 papers in peer-reviewed journals and more than 900 contributions in conferences. He has received many distinctions: the French Academic Palms, the Order of Merit, the Doctorate Honoris Causa at the University of Zaragoza in 2018 and the Silver medal from the French CNRS in 2019.

Elías Cueto

Elias Cueto is a professor of continuum mechanics at the University of Zaragoza. His research is devoted to the development of advanced numerical strategies for complex phenomena. In particular, in the last years, he has worked in model order reduction techniques and real-time simulation for computational surgery and augmented reality applications. His work has been recognized with the J. C. Simo award of the Spanish Society of Numerical Methods in Engineering; the European Scientific Association of Material Forming (ESAFORM) Scientific Prize; and the O.C. Zienkiewicz prize of the European Community on Computational Methods in Applied Sciences, ECCOMAS, among others. He is a fellow of the EAMBES society.

Benjamin Klusemann

Benjamin Klusemann is professor at the Leuphana University of Lüneburg as well as the head of the department "Solid State Materials Processing" at Helmholtz-Zentrum Hereon. He studied mechanical engineering followed by a PhD in computational mechanics at TU Dortmund. He worked as a postdoctoral researcher at RWTH Aachen University and the California Institute of Technology and as senior engineer at TU Hamburg. His current research interests include solid state processes, machine learning, micromechanics and process simulations, all with particular focus on experimental-modeling correlations. He received a number of awards, including the Richard-von-Mises-Prize of GAMM 2017, the ESAFORM scientific award 2019 and an ERC Consolidator Grant.

materials

Editorial

Empowering Materials Processing and Performance from Data and AI

Francisco Chinesta [1,2], **Elias Cueto** [3] and **Benjamin Klusemann** [4,5,*]

1   PIMM Laboratory & ESI Group Chair, Arts et Métiers Institute of Technology, CNRS, Cnam, HESAM Université, 151 Boulevard de l'Hôpital, 75013 Paris, France; francisco.chinesta@ensam.eu
2   ESI Group, Bâtiment Seville, 3bis Rue Saarinen, 50468 Rungis, France
3   Aragon Institute of Engineering Research, Universidad de Zaragoza, 50009 Zaragoza, Spain; ecueto@unizar.es
4   Institute of Product and Process Innovation, Leuphana University of Lüneburg, 21335 Lüneburg, Germany
5   Institute of Materials Mechanics, Helmholtz-Zentrum Hereon, 21502 Geesthacht, Germany
*   Correspondence: benjamin.klusemann@leuphana.de

Citation: Chinesta, F.; Cueto, E.; Klusemann, B. Empowering Materials Processing and Performance from Data and AI. *Materials* **2021**, *14*, 4409. https://doi.org/10.3390/ma14164409

Received: 14 July 2021
Accepted: 4 August 2021
Published: 6 August 2021

Publisher's Note: MDPI stays neutral with regard to jurisdictional claims in published maps and institutional affiliations.

Third millennium engineering is addressing new challenges in materials sciences and engineering. In particular, the advances in materials engineering, combined with the advances in data acquisition, processing and mining as well as artificial intelligence, allow new ways of thinking in designing new materials and products. Additionally, this gives rise to new paradigms in bridging raw material data and processing to the induced properties and performance. On the one hand, the linkage can be done purely on a data-driven basis, i.e., models are created from scratch based on the obtained experimental data alone, for instance with statistical methods or advanced methods of machine learning. Particularly obvious advantages of such kinds of models are that no simplification or assumptions need to be incorporated a priori, and that it allows real-time prediction, leading to a so-called digital twin of the specific material/process. However, such approaches typically face some general challenges, such as the necessity of (maybe unnecessarily) large and comprehensive datasets, because they rely only on the data themselves and allow prediction only within the investigated/trained dataspace. Another way of addressing the challenge of predicting the complex processing–structure–property relationships in materials is the enhancement of already existing physics-based models via data and machine learning tools, i.e., combining a physics-based model (often called virtual twin) and a data-based model, leading to a so-called hybrid twin [1]. In this regard, possible deviations of the physics-based model, which rely on a number of simplifications and assumptions, can be healed by correcting the model based on a data-driven approach, i.e., combining the advantages of both models.

This present topical issue is a compilation of contributions on novel ideas and concepts, addressing several key challenges using data and artificial intelligence:

- proposing new techniques for data generation and data mining;
- proposing new techniques for visualizing, classifying, modeling, extracting knowledge, explaining and certifying, data and data-driven models;
- processing data, for creating data-driven models from scratch when other models are absent, too complex or too poor for making valuable predictions;
- processing data for enhancing existing physic-based models to improve the quality of the prediction capabilities, and at the same time enabling data to be smarter.

Gong et al. [2] explored the material behavior of additive Ti-Mn samples, produced via the Laser Engineered Net Shaping (LENS™) technique and subsequent post-build heat treatment. In this work, novel cost-effective high-throughput experimental assays have been proposed and the feasibility of employing Gaussian process regression (GPR) for in-depth statistical analysis of these data has been shown. From the additive-manufactured compositionally graded cylinders and after heat treatment, samples were extracted for microstructural analysis and the mechanical properties were systematically determined via spherical indentation. The proposed approach allows to explore large material and process spaces. Via GPR, the process–microstructure–properties relationships were analyzed. The statistical approach showed a reliable and robust correlation between the Mn content and the specific ageing conditions to the microstructural parameters for the obtained data via the proposed high-throughput assays.

Huber [3] analyzed the microstructure–property relationships in nanoporous metals. In this study, a clear recommendation is given to always start with a dimensional analysis, ensuring a physically reasonable formulation of the problem at hand and a reduction of the problem to its minimum This included the incorporation of already existing physical knowledge by turning inputs and outputs into dimensionless parameters via physical normalization. The nanoporous microstructure was analyzed based on datasets obtained from finite element calculations of a beam model of the structure, investigated for macroscopic compression and nanoindentation. To analyze the microstructure–property relationships, not only dimensional analysis but also principal component analysis and visualization are used on the obtained datasets, where, in particular, machine learning algorithms are employed to identify key dependencies that allowed for deriving simple linear relationships in this case. It is revealed that the "scaling law of the work hardening rate has the same exponent as the Young's modulus" [3], an insight which can be transferred to the macroscopic behavior of such structures, allowing designing and tuning of such microstructures in terms of the specific objective.

Yun et al. [4] address the construction of models based on data involving complex microstructures. When addressing these complex microstructures, concise and complete morphological and topological descriptors are needed, as well as adequate metrics able to quantify the resemblance and proximity of different microstructures for which standard Euclidian distances do not perform correctly. For that purpose, Topological Data Analysis (TDA) is considered, and thus microstructures are represented from the so-called persistence images that contain most of the microstructural information while inheriting the appealing invariance topological features. With the persistence images defined in a vector space, usual metrics can be used to apply machine learning and data analytics techniques. Thus, the persistence images can be considered as inputs in machine learning-based nonlinear regressions, to infer the properties and performances associated with those complex microstructures. In particular, this work addresses the prediction of the effective thermal conductivity, inferred from the persistence image, without the need of further calculations.

For the identification and calibration of complex constitutive models, de Pablos et al. [5] presented an effective framework consisting of a two-stage procedure. In the first step, a metamodel is formulated and used to perform a Global Sensitivity Analysis, providing the sensitivity of each parameter in the selected material model with respect to the Quantities of Interest. Anisotropic Radial Basis Functions are used as kernel functions to build the linear metamodels in this work, allowing the generation of a large number of data points. The second step represent the parameter calibration for the most influential parameters based on Bayesian inference in combination with the application of Gaussian processes. This calibration process leads not only to the optimal mean value of the parameters, but it also provides information about their probability distributions. In this regard, the framework can help to select necessary subsequent experiments, but at the same time minimize the required number of experiments and numerical calculations. The application of the framework is illustrated on three well-known constitutive laws to describe elasto-viscoplastic material behavior.

Hartmaier [6] derived a novel mathematical formulation for a data-oriented constitutive model for elastic–plastic materials. In particular, this model consists of the identification of the yield function via a support vector characterization (SVC) algorithm to describe the elastic–plastic deformation. The model can be easily incorporated within finite element calculations. To reduce the dimensionality of the problem and without loss of generality, core physical principles are incorporated, i.e., working with the deviatoric stress as input data. The SVC algorithm can easily be trained, for instance, only based on the information whether a given stress state is within the elastic regime or not, leading to a machine learning yield function from which further information, such as its gradient, can be easily calculated. Overall, the presented data-oriented constitutive model shows a great flexibility in terms of application to all kind of anisotropy for elastic–plastic materials, where, for instance, the standard frameworks of continuum plasticity can still be employed within a finite element set-up. Identifying macroscopic yield functions based on microscopic calculations, the presented formulation is directly applicable as a kind of scale-bridging approach.

For scale bridging, Lu et al. [7] proposed a stochastic data-driven multilevel finite element (FE2) method, illustrated by the example of nonlinear electric conduction in graphene–polymer composites. At the microlevel, finite-element calculations of representative volume elements (RVE) of the graphene-polymer composites serve for the purpose of data generation. The data are used to train a neural network surrogate model, which relates the macroscopic electric field and volume fraction of graphene inclusions to the macroscopic electric flux. To reduce the necessary number of RVE calculations and at the same time to improve the accuracy of the surrogate model, a novel hybrid neural network–interpolation scheme is proposed. Since the surrogate model is computationally extremely cheap, it can efficiently be employed in a multilevel finite element approach. Additionally, considering certain properties as a stochastic field with given probabilistics at the macroscale, e.g., in the current study the volume fraction of heterogeneities, the surrogate model is used to identify a probabilistic model for each point at the macroscale, leading to a data-driven approach in a stochastic framework. Due to the computational efficiency, Monte Carlo simulations can be performed, allowing this approach to account for uncertainties at both scales.

Gonzalez et al. [8] deal with a particularly frequent problem in biomechanics: the large deviation of experimental results. In this framework, the standard deviation of the measurements may take a similar value to that of their mean value, thus making regression procedures a delicate issue. When this circumstance is present, they propose a combination of TDA and regression procedures operating on these just found data manifolds. To ensure the compliance to basic thermodynamic principles (conservation of energy, production of entropy), they suggest performing regression over a particularly interesting principle: the so-called General Equation for Non-Equilibrium Reversible-Irreversible Coupling (GENERIC). Generic is in fact a metriplectic formalism that, subject to adequate degeneracy conditions, satisfies by construction the first and second principles of thermodynamics. The resulting method thus ensures the thermodynamic correctness of the obtained constitutive laws, while dealing with the just-mentioned large deviations in experimental data.

Bock et al. [9] present a hybrid twin model for the use case of laser shock peening. The hybrid model consists of a physics-based model, i.e., a semi-analytical model, which is enhanced by a trained artificial neural network. The latter is accounting for the present deviation to a reference (high-fidelity) solution, which is determined by a finite element simulation of the laser shock peening process. On the one hand, the hybrid twin allows for a more efficient prediction of the residual stress field within the material in comparison to the high-fidelity model, since the computational costs are significantly reduced and similar to the ones required by the semi-analytical model. On the other hand, the hybrid twin outperforms a purely data-driven model by exhibiting enhanced generalization capabilities, i.e., predictions outside the process parameter region used for training, while requiring less data than the data-driven approach. The importance of a dimensionality analysis to identify the salient input features is emphasized, leading to low prediction errors.

Overall, the original articles compiled in this Research Topic give a taste of current ideas and concepts in how to enhance existing knowledge via new concepts of artificial intelligence to address current challenges in materials sciences and engineering. Such approaches provide new pathways, in particular in the research field of computational engineering, but with high impact on different engineering fields, such as material science and processing. At this point, as guest editors, we would like to thank all authors for their valuable contributions to this Special Issue!

Funding: This research received no external funding.

Conflicts of Interest: The authors declare no conflict of interest.

References

1. Chinesta, F.; Cueto, E.; Abisset-Chavanne, E.; Duval, J.L.; El Khaldi, F. Virtual, Digital and Hybrid Twins: A New Paradigm in Data-Based Engineering and Engineered Data. *Arch. Comput. Methods Eng.* **2020**, *27*, 105–134. [CrossRef]
2. Gong, X.; Yabansu, Y.C.; Collins, P.C.; Kalidindi, S.R. Evaluation of Ti–Mn Alloys for Additive Manufacturing Using High-Throughput Experimental Assays and Gaussian Process Regression. *Materials* **2020**, *13*, 4641. [CrossRef] [PubMed]
3. Huber, N. A Strategy for Dimensionality Reduction and Data Analysis Applied to Microstructure–Property Relationships of Nanoporous Metals. *Materials* **2021**, *14*, 1822. [CrossRef] [PubMed]
4. Yun, M.; Argerich, C.; Cueto, E.; Duval, J.L.; Chinesta, F. Nonlinear Regression Operating on Microstructures Described from Topological Data Analysis for the Real-Time Prediction of Effective Properties. *Materials* **2020**, *13*, 2335. [CrossRef] [PubMed]
5. de Pablos, J.L.; Menga, E.; Romero, I. A Methodology for the Statistical Calibration of Complex Constitutive Material Models: Application to Temperature-Dependent Elasto-Visco-Plastic Materials. *Materials* **2020**, *13*, 4402. [CrossRef] [PubMed]
6. Hartmaier, A. Data-Oriented Constitutive Modeling of Plasticity in Metals. *Materials* **2020**, *13*, 1600. [CrossRef] [PubMed]
7. Lu, X.; Yvonnet, J.; Papadopoulos, L.; Kalogeris, I.; Papadopoulos, V. A Stochastic FE2 Data-Driven Method for Nonlinear Multiscale Modeling. *Materials* **2021**, *14*, 2875. [CrossRef] [PubMed]
8. González, D.; García-González, A.; Chinesta, F.; Cueto, E. A Data-Driven Learning Method for Constitutive Modeling: Application to Vascular Hyperelastic Soft Tissues. *Materials* **2020**, *13*, 2319. [CrossRef] [PubMed]
9. Bock, F.E.; Keller, S.; Huber, N.; Klusemann, B. Hybrid Modelling by Machine Learning Corrections of Analytical Model Predictions towards High-Fidelity Simulation Solutions. *Materials* **2021**, *14*, 1883. [CrossRef] [PubMed]

 *materials* 

Article

Evaluation of Ti–Mn Alloys for Additive Manufacturing Using High-Throughput Experimental Assays and Gaussian Process Regression

Xinyi Gong [1], Yuksel C. Yabansu [2], Peter C. Collins [3] and Surya R. Kalidindi [1,2,*]

1   School of Materials Science and Engineering, Georgia Institute of Technology, Atlanta, GA 30332-0245, USA; xinyigong@outlook.com
2   George W. Woodruff School of Mechanical Engineering, Georgia Institute of Technology, Atlanta, GA 30332-0405, USA; canyabansu@gmail.com
3   Department of Materials Science and Engineering, Iowa State University, Ames, IA 50011, USA; pcollins@iastate.edu
*   Correspondence: surya.kalidindi@me.gatech.edu

Received: 16 September 2020; Accepted: 14 October 2020; Published: 17 October 2020

Abstract: Compositionally graded cylinders of Ti–Mn alloys were produced using the Laser Engineered Net Shaping (LENS™) technique, with Mn content varying from 0 to 12 wt.% along the cylinder axis. The cylinders were subjected to different post-build heat treatments to produce a large sample library of α–β microstructures. The microstructures in the sample library were studied using back-scattered electron (BSE) imaging in a scanning electron microscope (SEM), and their mechanical properties were evaluated using spherical indentation stress–strain protocols. These protocols revealed that the microstructures exhibited features with averaged chord lengths in the range of 0.17–1.78 µm, and beta content in the range of 20–83 vol.%. The estimated values of the Young's moduli and tensile yield strengths from spherical indentation were found to vary in the ranges of 97–130 GPa and 828–1864 MPa, respectively. The combined use of the LENS technique along with the spherical indentation protocols was found to facilitate the rapid exploration of material and process spaces. Analyses of the correlations between the process conditions, several key microstructural features, and the measured material properties were performed via Gaussian process regression (GPR). These data-driven statistical models provided valuable insights into the underlying correlations between these variables.

Keywords: high-throughput experimentation; additive manufacturing; Ti–Mn alloys; spherical indentation; statistical analysis; Gaussian process regression

1. Introduction

Modern metal additive manufacturing (AM) processes provide greatly expanded opportunities for producing engineered components possessing intricate geometries, novel material chemistries and internal structures. Furthermore, it is possible to tailor the material internal structures (hereafter simply referred to as microstructures) at different locations in the component both during the actual AM process and in subsequent (post-build) heat treatments in the effort to optimize its overall in-service functional performance [1–7]. Over the years, AM processes have been applied successfully in metal products with demonstrated benefits in net shaping, component repair, intricate geometry prototyping and component customization [8–15]. Although both experiments and physics-based multiscale material simulations have the potential to offer the data needed to gain insights into the correlations between the processing conditions and microstructure as well as the associated properties, our focus in this work was confined to experiments. This is mainly because the multiscale material models for

AM metal alloys are still largely under development [16–19]. The AM process usually consists of multiple steps, including substrate treatment, powder delivery, energy delivery, nozzle movement and post-build heat treatments [20–24]. Each of these process steps involves the selection of multiple parameters that could significantly influence the local thermal history and, thereby, the microstructure and associated properties. The central challenges encountered in the experimental exploration of the influence of AM processing conditions and material microstructure come from two main gaps in current capabilities in the field. First, there is a lack of validated high-throughput experimental assays that are cost-effective and require only small amounts of material in different conditions (e.g., a range of chemical compositions and process histories). Second, there is also a lack of established approaches capable of building reliable data-driven process–structure–property (PSP) linkages from limited data (i.e., a relatively small number of data points). This is particularly important for AM metal alloy development using experimental assays, because it is unlikely that one can accumulate the very large datasets needed by conventional machine learning approaches such as neural networks [25–28].

A number of different experimental protocols have been explored in recent literature for the rapid formulation of PSP linkages from experiments in metal AM [29–32]. The central challenges come from the need to prototype a large library of samples spanning the large ranges of chemistries and process histories relevant to metal AM, and subsequently characterizing their microstructures and mechanical properties. In this respect, it should be noted that AM inherently offers many advantages in prototyping libraries of samples with small volumes. Previous studies [32–35] have demonstrated the feasibility of manufacturing compositionally and functionally graded materials with microstructural gradients using a selective laser-melting technique. Similarly, Joseph et al. [36] demonstrated the feasibility of exploring the vast compositional space of high-entropy alloys (HEAs) using the direct laser-fabrication technique. Beyond the prototyping of the sample libraries, one also needs to address the challenges in the high-throughput characterization of the samples. It is important to note that the characterization should include both details of the material microstructures and their mechanical properties in order to meet our target of extracting PSP linkages that can accelerate material innovation for AM. In recent work, Saltzbrenner et al. [31] have demonstrated the viability of prototyping miniaturized tensile test specimens and conducting high-throughput tests in automated protocols. Although this approach has tremendous potential, in practice, it is often challenging to extract reliable and consistently reproducible mechanical properties because of the large heterogeneity exhibited by the AM samples. Since tensile testing requires a statistically homogeneous material condition in the entire gauge section for the successful evaluation of mechanical properties, any significant variation in the local thermal histories at different locations of the gauge section of the tensile test specimen can lead to a large variance in the values of the measured mechanical properties from such measurements. For AM samples, there is a critical need to explore other characterization methods that can evaluate mechanical properties in small material volumes without the need to make standardized tensile test samples.

In recent work [37], it was demonstrated that it is possible to prototype compositionally graded AM samples and characterize their mechanical properties using the stress–strain analysis protocols based on spherical indentation. This strategy appears to exhibit tremendous potential for the high-throughput extraction of the relationships between the processing conditions, microstructural features and properties [38,39]. Although this strategy has produced reliable data points, the size of the dataset (i.e., the number of data points obtained) is still rather small for the extraction of PSP linkages using emergent machine learning techniques. In this paper, we demonstrate novel workflows that extend significantly the previously demonstrated assays in multiple research directions: (i) the prototyping of a much larger library of AM Ti–Mn alloys employing intentionally induced compositional gradients coupled with different post-build heat treatments, and (ii) the use of data-driven model-building strategies such as Gaussian process regression (GPR) [40–46] for extracting practically useful correlations from experimental datasets. GPR offers many potential advantages compared to other machine learning approaches, including the ability to utilize smaller data sets (i.e., smaller numbers of data points) [42,44], rigorous treatment of uncertainty [47,48] and dynamic selection of new experiments that maximize the expected information gain [49–51]. This work explores and demonstrates a framework for high-throughput experimental assays to facilitate the efficient exploration of the AM process space as well as statistical analyses of the accumulated data using GPR approaches.

2. Methods

2.1. Experiments

Laser Engineered Net Shaping (known as LENS) [52–54] is the prototypical powder-blown direct energy deposition technique used for additive manufacturing. It often incorporates computer-controlled lasers as power sources and produces near-net shapes with sufficiently accurate dimensions as the final product [55] to eliminate the need for rough machining, making it popular in industry [52,53,56,57]. Due to characteristics such as its great reliability [53,54,58] and the low porosity [59–61] of the final products, LENS is widely employed in the customization and repair of intricate mechanical parts, including turbine blades [54,62–66]. The ability to control, independently, the powder flow from separate powder feeders in LENS allows for creating chemical gradients in the AM components [67–72]. This is of tremendous interest for the present study, which aims to prototype a large library of material samples of small volumes covering a range of alloy compositions and post-build heat treatments.

The binary Ti–xMn (x ranges from 0 to ~15 wt.% Mn) system was selected for this study. Titanium–manganese alloys are of great interest because of their numerous applications in aerospace, hydrogen storage and biomedical industries [73–75]. This range of manganese content in the alloy introduces a typical eutectoid β-stabilized system [73,76,77] that is notoriously susceptible to the segregation defect during solidification, known as "β-fleck" [78–81], and thus not suitable for traditional ways of developing cast/wrought titanium alloys. AM has the potential to eliminate β-fleck by taking advantage of the high thermal gradients and small molten pools, thereby reducing liquid-phase separation. By eliminating β-fleck, it may be possible to subsequently increase the strength through post-build aging heat treatments. An Optomec 750 LENS system was utilized to produce samples in this work. Elemental powders of Ti and Mn were introduced into the molten pool using two independently and automatically controlled powder feeders, one containing pure Ti and the other containing a mixture of elemental Ti and elemental Mn with a composition of 15 wt.% Mn. These powders, after leaving their powder feeders at preset feed rates, were mixed and focused into the molten pool by a multi-nozzle system. A Nd:YAG laser system producing near-infrared radiation with a wavelength of 1064 nm was focused coincident to the focal point of the powder, generating a local molten pool where melting and mixing occurred. The motion of the build plate was then controlled so that thin layers of controlled composition were deposited with predetermined width and thickness. The laser power at the molten pool was 410 W, and the nominal flow rate of the powders was ~2.6 g/min. The substrate travel speed (equivalent to the laser scan speed, but with a different reference frame) was 10 inch/min (the nonstandard units of inches and minutes are used when describing build parameters, as these are the standard units of the Optomec control system itself), and the hatch widths and layer thicknesses were 0.018 inch and 0.010 inch, respectively. The oxygen content in the glove box was maintained below 10 parts per million, with the balance being primarily argon gas.

Cylindrical samples with compositional gradients along their length were produced (see Figure 1a). Planning for the potential loss of volume of material due to cutting/machining/sample preparation (e.g., through the curfs of cuts), a small number of layers were programed to have the same composition at the beginning and end of the depositions. As a result, the samples produced for this study showed Mn content ranging from 0 to ~12 wt.% along the length. Three long strips (see the strip dimensions in Figure 1b) were sectioned out of the cylindrical sample and were subjected to different aging treatments. The aging treatments selected for the study included three different temperatures (500, 600 and 700 °C; see Figure 1c), while the aging time was kept the same, at four hours. The post-build aging treatment is expected to release residual stresses (these can be significant in the LENS technique due to the high power of the energy source, subsequent high temperature of the melt pool, fast cooling process and high build rate) as well as significantly alter the phase volume fractions and phase morphology, promoting possibilities of attaining improved properties.

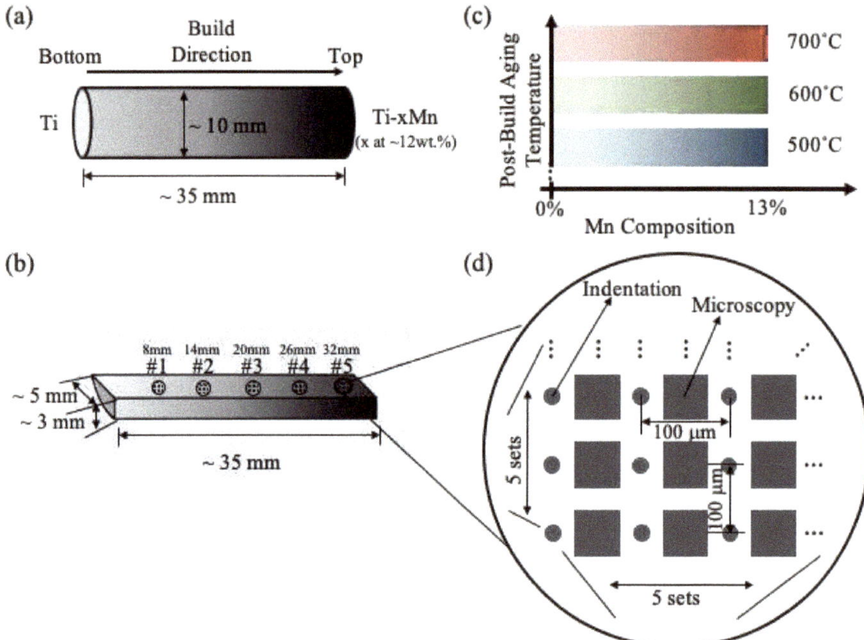

Figure 1. (**a**) Illustration of the layered Ti–Mn cylindrical sample manufactured by the Laser Engineered Net Shaping (LENS) process in this study. (**b**) Sample strip sectioned from (**a**) with compositional gradient along the length of the sample. Five locations were chosen longitudinally in each sample strip for characterization. They were 8, 14, 20, 26 and 32 mm away from the pure titanium end of the strip (labeled as #1–#5, respectively). (**c**) Three different sample strips were aged at three different temperatures (500, 600 and 700 °C, respectively) for four hours to produce the sample library used in this work. (**d**) A grid of indentation and microscopy characterization was performed at each location illustrated in (**b**). Each circle represents an indentation testing site, while the square represents the microscopy characterization site. Each measurement grid contained 5 by 5 indentation tests and the same number of microscopy characterizations. The test points in the grid were evenly spaced at 100 μm. Note all test sites shown in (**b**) are intentionally kept away from the thin end of the sample strips, making sure the sample has at least 2 mm thickness at the indentation test sites.

After aging, all the sample strips were prepared for microscopy and spherical indentation stress–strain measurements using standard metallography protocols established previously for titanium alloys [82]. These included grinding (P240 and P1200 SiC papers), followed by polishing steps with decreasing abrasive particle sizes (9, 3 and 1 μm diamond suspensions), while making sure every step removed the surface deformation introduced by the previous step. A solution of 0.06 μm colloidal silica suspension with hydrogen peroxide in the ratio of 5 to 1 was employed in a final polishing step to produce the desired surfaces for microscopy and indentation.

The main focus of this study is exploring high-throughput experimental assays for exploring large material spaces for AM. Five locations were selected longitudinally in each sample strip (see Figure 1b) for microstructure characterization and indentation tests. The transverse directions on the sample surface are not expected to exhibit any significant compositional gradients. Multiple indentation measurements and back-scattered electron (BSE) imaging were performed on a 5 × 5 grid at each of the five selected locations (illustrated in Figure 1d). Indentation tests were performed on an Agilent G200 (Santa Clara, CA, USA) with a continuous stiffness measurement (CSM) under a constant strain rate of 0.05/s and 800 nm indentation depth. The CSM was set at a 45 Hz oscillation with a 2 nm displacement amplitude [83]. A Tescan Mira XMH field emission SEM (Warrendale, PA, USA) with a 20 kV accelerating voltage was used to capture back-scattered electron (BSE) images. Energy dispersive spectroscopy (EDS) was performed at the five locations shown in Figure 1b to measure the Mn content. At each location, five EDS measurements randomly distributed within the 5×5 grid (established in Figure 1d) were performed. Each measurement was carried out by first mapping the element concentration distribution of a 50 μm × 50 μm area and then calculating the average element composition according to the map. A Hitachi SU8230 SEM (Tokyo, Japan) equipped with Oxford EDAX and Aztec analysis software was used for EDS analysis. Beam calibrations with a 100% copper plate were used for EDS quantification. The accelerating voltage was kept at 20 kV and beam intensity at 20 μA for these measurements.

2.2. Microstructure Analysis and Quantification

The two-phase BSE images were segmented with Otsu's method [84,85]. Otsu's method separates the intensity distribution of an image into two classes by using a threshold. The threshold value is determined to maximize the interclass variance (or minimize the intraclass variance). Otsu's thresholding was performed using the "graythresh" function of the numerical computing software MATLAB [86]. The segmented (binary) images were used to compute the volume fraction of the β phase. Additionally, averaged chord lengths (CL) [87,88] were computed to quantify the length scales of the α and β phase regions in the microstructure. The procedures used to identify the chords are based on pixelized representations of the images and have been described in prior work [87,89]. A chord is defined as a line segment (measured as the number of pixels) that completely lies inside a distinct material phase, whose extension in any direction by even one pixel encounters pixels of a different material phase.

2.3. Mechanical Characterization

The spherical indentation stress–strain protocols [90–92] employed in this study are built largely on Hertz's theory [93,94] for elastic frictionless contact between two isotropic bodies with parabolic surfaces (see Figure 2a). The relevant relationships are summarized below:

$$P = \frac{4}{3}E_{eff}R_{eff}^{1/2}h_e^{3/2} \tag{1}$$

$$a = \sqrt{R_{eff}h_e} = \frac{S}{2E_{eff}} \tag{2}$$

$$\frac{1}{E_{eff}} = \frac{1 - v_i^2}{E_i} + \frac{1 - v_s^2}{E_s} \tag{3}$$

$$\frac{1}{R_{eff}} = \frac{1}{R_i} + \frac{1}{R_s} \tag{4}$$

where P and h_e denote the indentation load and elastic indentation displacement, E_{eff} and R_{eff} denote the effective modulus and the radius of the indenter-sample system, subscripts i and s correspond to the indenter and the sample, and the Young's modulus and Poisson's ratio are denoted as E and v. In Equation (2), S ($= dP/dh_e$) denotes the elastic stiffness (also known as the harmonic stiffness in continuous stiffness measurement (CSM) protocols [83,95,96]). Building on these relationships, one can define the indentation stress, σ_{ind}, and the total indentation strain (includes the elastic and plastic strains), ε_{ind}, as

$$\sigma_{ind} = \frac{P}{\pi a^2} \tag{5}$$

$$\varepsilon_{ind} = \frac{4}{3\pi}\frac{h_s}{a} \tag{6}$$

where h_s is the corrected sample displacement (subtracting the displacement in the indenter, h_i, from the total displacement, h) and is computed using

$$h_s = h - \frac{3\left(1 - v_i^2\right)P}{4E_i a} \tag{7}$$

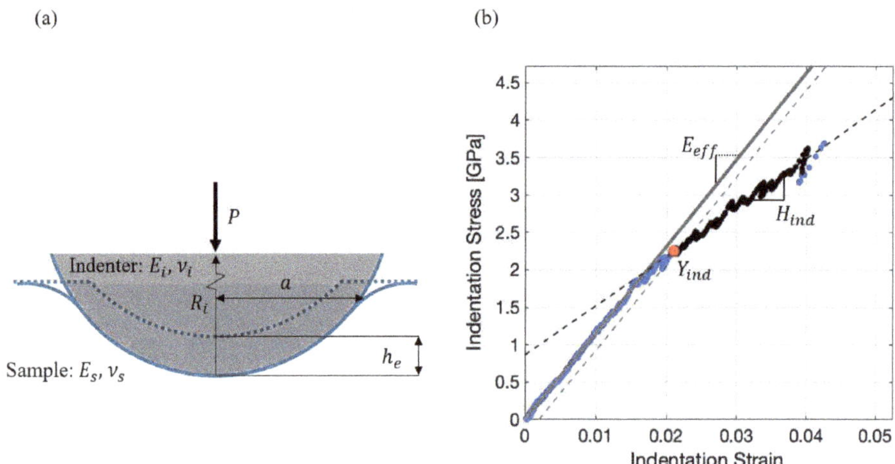

Figure 2. (**a**) Illustration of spherical indentation. (**b**) Indentation stress–strain curve acquired from Location #4 (see Figure 1b) of the strip heat treated at 700 °C. The slope illustrated in the elastic portion of the indentation stress–strain curve is the effective modulus, E_{eff}. The red dot represents the indentation yield strength Y_{ind} corresponding to a 0.002 offset indentation plastic strain, while the black segment (from 0.005 to 0.02 in offset indentation plastic strain) represents the data used to estimate the indentation work hardening rate H_{ind}.

The indentation stress and indentation strain defined in Equations (5) and (6) exhibit a linear relationship in purely elastic indentations, where the slope of the indentation stress–strain curve is referred to as the indentation modulus, E_{ind}. For an isotropic material, the indentation modulus and the Young's modulus are related as

$$E_{ind} = \frac{E_s}{\left(1 - v_s^2\right)} \tag{8}$$

On the indentation stress–strain curve (see Figure 2b), a 0.2% offset indentation plastic strain is used to define the indentation yield strength, Y_{ind}. The indentation stress–strain curve between the 0.5% and 2% offset indentation plastic strains is fitted with a linear regression [82,97] to compute the indentation work hardening rate, H_{ind}. In prior work [82,91,98,99], Equations (5) and (6) were demonstrated to produce meaningful elastic–plastic indentation stress–strain curves that show an elastic–plastic regime following an initial elastic regime (see Figure 2b).

In the present study, spherical indentations on a 5×5 grid were performed with a uniform spacing of 100 μm (see Figure 1d). A diamond indenter tip with a nominal radius of 100 μm was used in all the tests reported in this work. Each indentation produced a contact area of about 150 μm² (contact radius of roughly 7 μm) at indentation yield and hence reflected the effective response of the two-phase microstructures obtained in the sample library (micrographs presented later). The spacing between indentations was designed to be large enough to minimize the interference between neighboring indentations. However, it was also important to keep the spacing small enough so that the compositional variation between the indentation locations within each grid was very small.

2.4. Gaussian Process Modeling

In this study, Gaussian process regression (GPR) was employed to establish quantitative correlations between various measured quantities of interest in the extracted dataset. GPR is a nonparametric machine learning method that employs joint probability distributions to the available training data (usually a small dataset) in order to make probabilistic predictions for new inputs. This is accomplished by treating the correlations as a Gaussian process (GP) defined by only its mean and variance. Let $t = \{t_1, t_2, \ldots, t_N\}$ and $y = \{y_1, y_2, \ldots, y_N\}$ denote the target and prediction, respectively, where N denotes the number of training points. Then, the relationship between the target t and prediction y can be written as

$$t = y + \varepsilon \tag{9}$$

where ε is a column vector containing the residuals of N observations. A GP governing the joint distribution between the predictions can be written as

$$y(x) \sim \mathcal{N}\left(\mu(x), K\left(x, x'\right)\right) \tag{10}$$

where x denotes a $1 \times D$ input vector, and $\mu(x)$ and $K\left(x, x'\right)$ represent the mean and the covariance of the GP, respectively. In Equation (10), $\mathcal{N}()$ denotes a multivariate Gaussian distribution.

The covariance of the GP is generally computed using a kernel function $k\left(x, x'\right)$. In the present study, the %Mn and the post-heat treatment temperature are treated as the two (i.e., $D = 2$) independent variables (i.e., inputs) for the model-building effort in this study. The outputs for the study will include a number of microstructure statistics as well as the measured mechanical properties. The automatic relevance determination squared exponential (ARDSE) [100–102] was selected as the kernel for computing the covariance matrices. This ARDSE kernel is mathematically expressed as

$$k\left(x, x'\right) = \sigma_f^2 exp\left(-\frac{1}{2}\left(\frac{(x_T - x'_T)^2}{l_T^2} + \frac{(x_c - x'_c)^2}{l_c^2}\right)\right) + \sigma_n^2 \delta_{xx'} \tag{11}$$

where σ_f, l_T, l_c and σ_n are the hyperparameters that control the fidelity of the GP model, and the subscripts T and c refer to the two input variables (i.e., the post-heat treatment temperature and %Mn). The hyperparameters in the kernel provide more valuable information about the trends and relationships between the inputs and the outputs, especially when compared to conventional correlation techniques such as the Pearson correlation coefficient [103]. More specifically:

(1) σ_f is called the output scaling factor and determines the variance of the output values. A higher value of σ_f indicates that the values of the output are widely spread. The ratio of σ_f to the output noise σ_n (discussed later) determines the uncertainty of the predictions made from the GP model.

(2) l_T and l_c are the interpolation length scale parameters associated with the two input variables and capture the sensitivity of the output variable to the changes in the respective input values. Lower length scale values exhibit shorter memory, leading to sharper fluctuations and more complex nonlinear mapping between the inputs and the output. In other words, lower values of the interpolation length parameter indicate a higher sensitivity of the output to the input value (for the selected input variable). Conversely, larger values of the interpolation length parameters indicate low levels of correlation between the output and the corresponding input variable.

(3) σ_n is called the output noise hyperparameter and captures the variance in the training data. For the present study, where the training data are obtained from experiments, this variance can arise from variations in the execution of the experimental assays themselves or variations in the

application of the analysis protocols (e.g., image segmentation). σ_n is assumed to be the same for the entire input domain (also called homoscedasticity [104]).

The hyperparameters in Equation (11) are generally optimized to produce the most reliable predictions for test data points. For this, one must formulate a conditional distribution of test points, t^*, given the evidence of training points, t. Let the train and test datapoints be represented by matrices X and X^* of sizes $N \times D$ and $N^* \times D$, respectively, where N^* reflects the number of test points. The overall covariance matrix can be partitioned as

$$C = \begin{bmatrix} K(X,X) & k^*(X,X^*) \\ k^{*\dagger}(X,X^*) & K^*(X,X) \end{bmatrix} \tag{12}$$

where $\dagger$ represents the transpose. Each term of the covariance matrix in Equation (12) is computed using the kernel function from Equation (11). The predictive distributions for the test points, given the training points, can be expressed as [100,101].

$$\mu^* = k^{*\dagger}K^{-1}t$$
$$\Sigma^* = K^* - k^{*\dagger}K^{-1}k^* \tag{13}$$

where μ^* and Σ^* denote the prediction means and variances (i.e., uncertainty), respectively, for the test points. The central challenge in the computations described in Equation (13) comes from the need to perform an inverse on the $N \times N$ covariance matrix of the training points, which requires $O(N^3)$ computations. Once K^{-1} is obtained, predictions for the test points can be realized through standard matrix multiplication/addition operations, which require only $O(N^2)$ computations [100,101]. Note, also, that in the applications explored in this work, the number of data points is quite small. Therefore, the one-time computational cost of the inversion operation in Equation (13) does not represent a major challenge for the present study.

3. Results

Figure 3a shows a typical BSE micrograph taken from a location approximately 14 mm away from the pure titanium end of the sample strip aged at 500 °C. In this micrograph, the lamellated hcp (hexagonal closest packing) α-Ti and bcc (body-centered cubic) β-Ti are visible as the darker and brighter regions, respectively. The EDS measurements also show less than 1 wt.% of manganese for the darker phase and about 15 wt.% of manganese for the brighter phase, thereby identifying these regions as α and β titanium, respectively. A map sum spectrum was also taken, measuring the average manganese content at 5.8 wt.% for this scan. Similar measurements were carried out at each location identified in Figure 1b for each sample strip (i.e., each composition–post-heat treatment combination). The results are presented in Figure 3b. As expected, it is seen that the variation of the Mn content along the strip is highly consistent between the different strips. The manganese composition rises from the pure titanium end but peaks at about 20 mm and stays at about 12 wt.%. Note that this 12 wt.% Mn is a little lower than the target composition of 15 wt.%. The maximum manganese composition is present over a few millimeters at the end of the build, as programmed into the original motion control source code. The deviations between the obtained local composition and the targeted composition are attributed to variations in the local elemental powder being fed or, as is the case here, due to an intentional extension of a region with the maximum Mn concentration. Since our primary interest in the present study is the development of a framework for establishing the correlations between the processing parameters, microstructure statistics and the properties, we have not iterated with different starting powder mixtures to attain specific compositions in the produced samples. Instead, our focus will be on the protocols needed to acquire, efficiently, the material data needed for the targeted correlations.

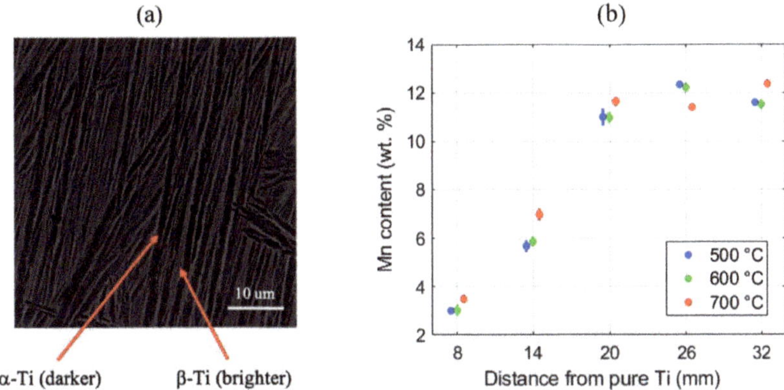

Figure 3. (**a**) Back-scattered electron (BSE)-SEM image for the sample strip aged at 500 °C for four hours and at the location where the Mn content was 5.8 wt.%. It depicts the dual-phase microstructure of the sample, where the darker phase is α-Ti and the brighter phase is β-Ti. (**b**) Means and standard deviations from the energy dispersive spectroscopy (EDS) measurements of the Mn content at the five locations for all three high-throughput (HT) sample strips produced for this study. For clarity, all 500 °C and 700 °C values are intentionally shifted slightly in the negative and positive x directions, respectively. All points in each group correspond to the same nominal distance indicated by the axis ticks.

Multiple BSE micrographs were obtained corresponding to each combination of manganese composition and post-heat treatment temperature. The volume fractions estimated from the segmented images are shown in Figure 4. It is seen that the β volume fraction increased with Mn content and with the temperature of the post-build aging treatments. This is because post-build aging at a higher temperature pushes the microstructure to be close to its equilibrium state. Note that the high manganese locations subjected to the low 500 °C treatment (see the bottom left micrograph in Figure 5) produced a small-scale (10–100 nm) secondary α phase [33] in addition to the bigger (~2 μm) primary α laths. Such secondary α is expected, especially at these lower temperatures, and results when new nucleation events become more favorable and accelerate the rate of transformations.

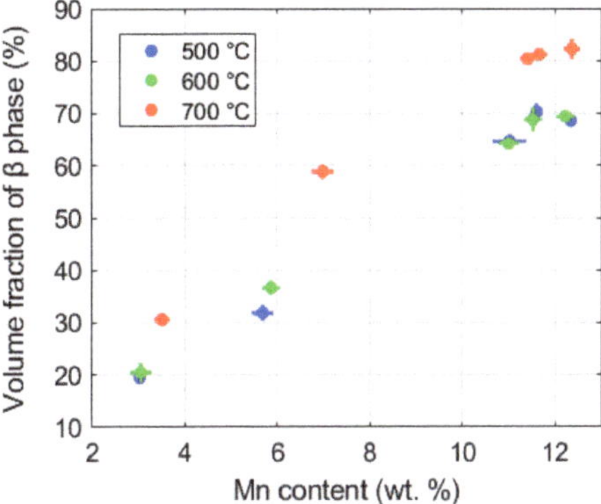

Figure 4. Means and standard deviations of the percentage volume fractions of the β phase obtained for the different Mn contents and post-build aging heat treatments.

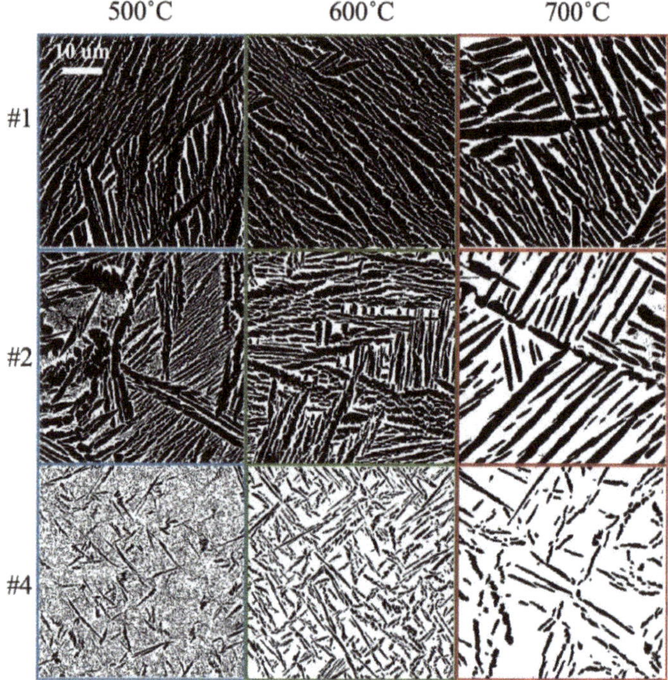

Figure 5. Segmented SEM-BSE images for the sample library produced and studied in this work. The left, middle and right columns correspond to aging heat treatments of 500, 600 and 700 °C, respectively. The rows correspond to different locations exhibiting different manganese compositions (see Figures 1b and 3b). The black phase in these micrographs represents α-Ti, while the white phase represents β-Ti.

For the computation of the averaged CLs, all the chords in the micrograph were collected at intervals of 2.5 degrees to avoid imaging orientation bias. The averaged value of all the collected chord lengths for each phase at each of the five sample locations identified is reported in Figure 6. The averaged CL of the dominant β phase decreased consistently with an increase in the Mn content. By contrast, the averaged chord length of the β phase increased with a higher manganese content, with the higher aging temperature promoting a more drastic change.

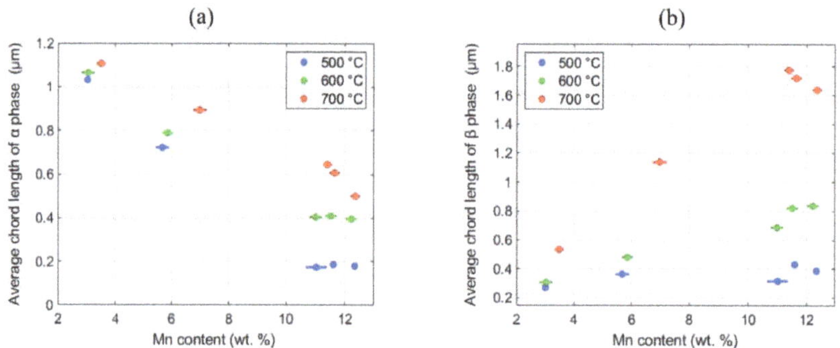

Figure 6. Averaged chord lengths (CLs) of (**a**) α phase and (**b**) β phase at the selected five locations for all three high-throughput (HT) sample strips studied in this work.

Spherical indentation tests were performed on a grid of twenty-five sites for each of the five sample locations (see Figure 1b,d) for all three sample strips. Figure 7 summarizes the measured values of elastic moduli, indentation yield strengths and indentation hardening rates at each of the five locations on all three strips studied in this work. It is observed that the measured indentation moduli did not show significant variations between different locations and between different sample strips. On the other hand, a strong positive correlation was observed between the Mn content (which also correlated well with the beta volume fraction (see Figure 4)) and the indentation yield strength as well as the indentation hardening rate.

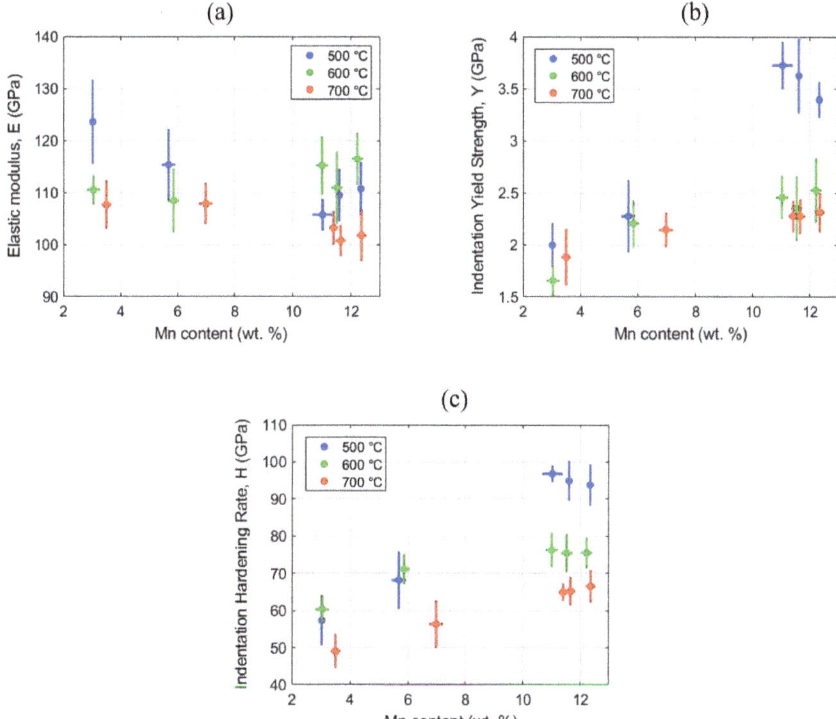

Figure 7. Mechanical properties estimated from the spherical indentation stress–strain protocols: (a) Young's modulus, (b) indentation yield strength and (c) indentation initial hardening rate. The blue, green and red boxes correspond to the 500, 600 and 700 °C aged strips, respectively.

It is clearly seen that the various microstructural features (β volume fraction and the averaged CLs of the α and β regions) and the resulting mechanical properties are highly correlated with each other. In order to analyze the effects of the process conditions (i.e., the aging temperature and Mn content) on the microstructural features and the resulting mechanical properties, it is necessary to conduct a statistical analysis. Gaussian process regression (GPR) was employed in this work for this purpose. As mentioned before, the hyperparameters of the kernel function provide reliable insights into the sensitivities of the different inputs to the outputs of interest.

A separate GP was built for each of the six outputs listed in Table 1, while using the post-build aging temperature and Mn content as features (i.e., independent variables). Traditionally, GP models are built to provide predictions for new inputs. However, in the present application, the size of the dataset is too small to formally establish a reliable predictive model with rigorous cross-validation. Therefore, it was decided to use the GP models to provide reliable insights into the sensitivities between the various measured quantities in this study. The interpolation length scale parameters established by these GP models and summarized in Table 1 are ideally suited for extracting such insights. As a specific example, it is seen that the interpolation length scale hyperparameter for the aging temperature in the GP model for the averaged CL for the α phase is very large, especially compared to the corresponding values obtained for the GP models for the other five outputs. This indicates a much lower sensitivity of the averaged CL of the α phase to the aging temperature. In fact, the averaged β-CL and the indentation hardening rate are found to exhibit the highest levels of sensitivity to the aging temperature. The table also indicates that all the microstructural parameters exhibited strong sensitivity to the Mn content, with the β volume fraction showing the highest sensitivity.

Table 1. Gaussian process regression (GPR) interpolation length hyperparameters and the mean absolute percentage error (MAPE) for each of the six outputs selected for these models. CL denotes the averaged chord length, VF is the volume fraction, Y is the indentation yield strength, and E is the Young's modulus.

GPR Results	CL α	CL β	VF-β	Y	H	E
l_T	667.17	141.85	263.23	199.12	143.38	211.78
l_c	10.95	10.48	9.39	16.83	14.78	12.49
σ_f	19.52	18.47	0.55	2.48	53.97	92.17
σ_n	0.93	1.66	0.01	0.17	2.64	1.39
σ_f/σ_n	20.88	11.13	45.59	14.50	20.41	66.10
$MAPE$	6.89	9.87	3.54	6.26	3.32	2.02

A comparison of the output scaling factor σ_f, which controls the overall spread of the output values in the entire dataset with the output noise parameter σ_n, provides insight into the combined overall predictive capability of the GPR model. The ratio σ_f/σ_n is referred to as the output-to-noise ratio and reflects the capability of the selected inputs in influencing the predicted output. For example, a very high value of σ_f/σ_n obtained in a specific GPR model indicates that the selected inputs (i.e., the Mn content and the aging temperature) are able to reliably account for most of the observed variations in the selected output in the collected dataset. In other words, the GP models with high values of σ_f/σ_n are indeed more mature and can be used reliably in making predictions for new inputs. In Table 1, it is seen that the GPR models for the elastic modulus and the β volume fraction show very high values of σ_f/σ_n, indicating that these models are able to account for almost all of the measured variations in these quantities in the data aggregated in this work. Similarly, a low value of σ_f/σ_n might suggest a lack of adequate correlations between the selected inputs and the output. This could suggest that there is inherently more noise in the measured values of the selected output, the possible existence of as-yet-unidentified inputs influencing the output variable, or both. In Table 1, the lowest value of σ_f/σ_n was obtained for the averaged CL for the β regions. In this study, we believe this is because of the inherent noise resulting from the protocols used to estimate this attribute from the micrographs (i.e., the segmentation and CL protocols). In other words, if one intends to establish more accurate correlations for the averaged CL for the β regions, it would be prudent to improve the protocols used to extract this value.

Table 1 also summarizes the mean absolute percentage error (MAPE) using a leave-one-out cross-validation strategy. This entails obtaining a model by setting aside one data point at a time in establishing the GPR model and subsequently testing the obtained model on the excluded point. The process is then systematically repeated for all available data points, and the MAPE is computed based on the obtained errors. It is seen from Table 1 that the GPR model for the elastic modulus exhibits the highest accuracy, while the GPR models for the averaged CL for the β regions exhibited the lowest accuracy. It is also seen that this is consistent with the σ_f/σ_n values.

4. Discussion

Compared to the conventional assays, the high-throughput assays employed in this work required significantly smaller material volumes. It should be noted that a total of 15 material conditions (obtained by combining three different post-build aging heat treatments with five different compositions) were produced and studied with relatively low overall effort and cost. In fact, the high-throughput (HT) assays described in this work exhibit tremendous potential for further scale-up, allowing the rapid evaluation of several hundreds of material conditions. In addition to requiring only small volumes of the material, the time and effort needed for the proposed HT assays are also significantly lower compared to those for the conventional assays. This is because the sample preparation steps only require standard metallography protocols that are needed anyway if the material microstructure is to be documented in such explorations.

As demonstrated in previous studies [37,39,82,105], the averaged values from the multiple indentations summarized in Figure 7 provide reliable measures of the bulk properties measured in standardized tests. The estimated Young's moduli did not show clearly identified trends between the different material conditions explored in this study, and fell in the range of 97–130 GPa. In general, one might expect a decrease in the Young's modulus with an increase in the β volume fraction, as the β phase is expected to exhibit a lower elastic modulus compared to the α phase [106,107]. Although one might be tempted to infer such a trend from Figure 7, it is not clearly evident, as the noise in the measurements is of the same order as the overall variation among the tested samples. However, the overall range of the values estimated in this work is comparable to the ranges published in the literature [108,109] for similar compositions.

As seen in Figure 7, the indentation yield strengths and the indentation hardening rates in the early stages of the imposed plastic deformation increased systematically with the increase in Mn content. There is, as expected, a clear positive correlation between the indentation yield strength, the indentation hardening rates and the Mn content. Strengthening due to secondary phase, solid solution strengthening, and grain boundary strengthening are likely to contribute to the observed increase in the indentation strength with the increase in Mn content. Based on prior work [110], the indentation yield strengths can be converted to tensile yield strengths using a scaling factor of 2.0. Using this scaling factor, the tensile yield strengths for the material conditions studied are expected to range between 828 and 1864 MPa, which is among the highest ranges reported [108,111–113] for similar compositions. Interestingly, the highest values were obtained for the samples with the highest Mn content and the lowest post-build aging treatment. Indeed, the corresponding microstructures also showed a relatively high β volume fraction of about 65% and highly refined microstructures with averaged CLs of 0.17 and 0.32 µm in the α and β phases, respectively (see Figure 5). The refined length scales are also responsible for the high indentation hardening rates measured in our experiments, because of the presence of a much larger number of interfaces per unit volume of the material. The fact that our high-throughput protocols easily identified the viability of obtaining a very high yield strength in the Ti–Mn alloys along with the features identified in their microstructures clearly testifies to the unique benefits of our approach for the rapid exploration of large material spaces.

5. Conclusions

Novel high-throughput assays have been proposed and demonstrated to rapidly explore large material spaces reflecting the many combinatorial selections in material compositions and AM process parameters such as post-build aging treatments. More specifically, this study successfully conducted such an evaluation using Ti–Mn alloy systems processed by LENS, which allowed for the generation of samples with controlled composition gradients. Combining this strategy with spherical indentation stress–strain protocols allowed for a rapid exploration of the mechanical properties of the produced samples in small material volumes. Most importantly, this rapid exploration revealed that a Mn content of about 12% with a post-build heat treatment of 500 °C produced an unusually hard material with an expected tensile yield strength of 1864 MPa. The dataset generated in this study was analyzed rigorously using GPR models. The use of these statistical approaches revealed that the use of the Mn content and the post-build aging treatment as inputs does lead to reliable correlations with microstructure measures such as the β volume fraction and the averaged CLs of the α and β regions, as well as their mechanical properties such as the Young's modulus, indentation yield strength and indentation hardening rate. These correlations revealed the relative sensitivities of the different outputs to the selected inputs as well as the high levels of inherent noise in the estimation of the averaged CLs of β regions. The GPR models built with the limited data obtained in this work showed reasonable accuracy in leave-one-out cross-validation. This study established the feasibility and value of employing GPR approaches in the rigorous statistical analyses of the datasets produced in the proposed high-throughput assays for material exploration.

Author Contributions: Conceptualization, X.G., Y.C.Y., P.C.C. and S.R.K.; methodology, X.G., Y.C.Y., P.C.C. and S.R.K.; software, X.G. and Y.C.Y.; validation, X.G. and Y.C.Y.; formal analysis, X.G., Y.C.Y. and S.R.K.; investigation, X.G., Y.C.Y. and S.R.K.; resources, P.C.C. and S.R.K.; data curation, X.G. and Y.C.Y.; writing of the original draft preparation, X.G., Y.C.Y., P.C.C. and S.R.K.; writing of review and editing, X.G., Y.C.Y., P.C.C. and S.R.K.; visualization, X.G., Y.C.Y., P.C.C. and S.R.K.; supervision, P.C.C. and S.R.K.; project administration, P.C.C. and S.R.K.; funding acquisition, P.C.C. and S.R.K. All authors have read and agreed to the published version of the manuscript.

Funding: X.G., P.C.C. and S.R.K. wish to acknowledge support provided by the National Science Foundation awards 1435237 and 1606567 ("DMREF/Collaborative Research: Collaboration to Accelerate the Discovery of New Alloys for Additive Manufacturing") with gratitude. Y.C.Y. gratefully acknowledges funding from N00014-18-1-2879. P.C.C. would also like to gratefully acknowledge the facilities at the University of North Texas, as well as Chandana Avasarala and Peyman Samimi for their contributions in the deposition of this Ti–xMn specimen.

Conflicts of Interest: The authors declare no conflict of interest.

References

1. Shamsaei, N.; Yadollahi, A.; Bian, L.; Thompson, S.M. An overview of Direct Laser Deposition for additive manufacturing; Part II: Mechanical behavior, process parameter optimization and control. *Addit. Manuf.* **2015**, *8*, 12–35. [CrossRef]

2. Ramirez, D.; Murr, L.; Martinez, E.; Hernandez, D.; Martinez, J.; Machado, B.; Medina, F.; Frigola, P.; Wicker, R. Novel precipitate–microstructural architecture developed in the fabrication of solid copper components by additive manufacturing using electron beam melting. *Acta Mater.* **2011**, *59*, 4088–4099. [CrossRef]

3. Imanian, A.; Leung, K.; Iyyer, N.; Li, P.; Warner, D.H. Optimize Additive Manufacturing Post-Build Heat Treatment and Hot Iso-Static Pressing Process Using an Integrated Computational Materials Engineering Framework. *ASME 2018 Int. Mech. Eng. Congr. Expo.* **2018**, *52019*, 00202064. [CrossRef]

4. Garner, E.; Kolken, H.M.; Wang, C.C.; Zadpoor, A.A.; Wu, J. Compatibility in microstructural optimization for additive manufacturing. *Addit. Manuf.* **2019**, *26*, 65–75. [CrossRef]

5. Wang, Y.; Zhang, L.; Daynes, S.; Zhang, H.; Feih, S.; Wang, M.Y. Design of graded lattice structure with optimized mesostructures for additive manufacturing. *Mater. Des.* **2018**, *142*, 114–123. [CrossRef]

6. Doubrovski, Z.; Verlinden, J.C.; Geraedts, J.M. Optimal design for additive manufacturing: Opportunities and challenges. In Proceedings of the ASME 2011 International Design Engineering Technical Conferences and Computers and Information in Engineering Conference, Washington, DC, USA, 28–31 August 2011; pp. 635–646. [CrossRef]

7. Gu, D. *Laser Additive Manufacturing of High-Performance Materials*; Springer: Berlin/Heidelberg, Germany, 2015.

8. Guo, N.; Leu, M.C. Additive manufacturing: Technology, applications and research needs. *Front. Mech. Eng.* **2013**, *8*, 215–243. [CrossRef]

9. Zhai, Y.; Lados, D.A.; LaGoy, J.L. Additive manufacturing: Making imagination the major limitation. *JOM* **2014**, *66*, 808–816. [CrossRef]

10. Herderick, E. Additive manufacturing of metals: A review. In Proceedings of the Materials Science & Technology Conference and Exhibition 2011 (MS & T 11), Columbus, OH, USA, 16–20 October 2011; pp. 1413–1425.

11. Vaezi, M.; Chianrabutra, S.; Mellor, B.; Yang, S. Multiple material additive manufacturing–Part 1: A review: This review paper covers a decade of research on multiple material additive manufacturing technologies which can produce complex geometry parts with different materials. *Virtual Phys. Prototyp.* **2013**, *8*, 19–50. [CrossRef]

12. Seepersad, C.C. Challenges and opportunities in design for additive manufacturing. *3D Print. Addit. Manuf.* **2014**, *1*, 10–13. [CrossRef]

13. Reeves, P.; Tuck, C.; Hague, R. Additive manufacturing for mass customization. In *Mass Customization*; Springer: London, UK, 2011; pp. 275–289. [CrossRef]

14. Attaran, M. The rise of 3-D printing: The advantages of additive manufacturing over traditional manufacturing. *Bus. Horiz.* **2017**, *60*, 677–688. [CrossRef]

15. Paoletti, I. Mass customization with additive manufacturing: New perspectives for multi performative building components in architecture. *Procedia Eng.* **2017**, *180*, 1150–1159. [CrossRef]

16. Gu, D.; Ma, C.; Xia, M.; Dai, D.; Shi, Q. A multiscale understanding of the thermodynamic and kinetic mechanisms of laser additive manufacturing. *Engineering* **2017**, *3*, 675–684. [CrossRef]

17. Markl, M.; Körner, C. Multiscale modeling of powder bed–based additive manufacturing. *Annu. Rev. Mater. Res.* **2016**, *46*, 93–123. [CrossRef]

18. Li, C.; Fu, C.; Guo, Y.; Fang, F. A multiscale modeling approach for fast prediction of part distortion in selective laser melting. *J. Mater. Process. Technol.* **2016**, *229*, 703–712. [CrossRef]

19. Francois, M.M.; Sun, A.; King, W.E.; Henson, N.J.; Tourret, D.; Bronkhorst, C.A.; Carlson, N.N.; Newman, C.K.; Haut, T.S.; Bakosi, J. Modeling of additive manufacturing processes for metals: Challenges and opportunities. *Curr. Opin. Solid State Mater. Sci.* **2017**, *21*. [CrossRef]

20. Frazier, W.E. Metal additive manufacturing: A review. *J. Mater. Eng. Perform.* **2014**, *23*, 1917–1928. [CrossRef]

21. Lewandowski, J.J.; Seifi, M. Metal additive manufacturing: A review of mechanical properties. *Annu. Rev. Mater. Res.* **2016**, *46*, 151–186. [CrossRef]

22. Kok, Y.; Tan, X.P.; Wang, P.; Nai, M.; Loh, N.H.; Liu, E.; Tor, S.B. Anisotropy and heterogeneity of microstructure and mechanical properties in metal additive manufacturing: A critical review. *Mater. Des.* **2018**, *139*, 565–586. [CrossRef]

23. Bhavar, V.; Kattire, P.; Patil, V.; Khot, S.; Gujar, K.; Singh, R. A review on powder bed fusion technology of metal additive manufacturing. In *Additive Manufacturing Handbook*; CRC Press: Boca Raton, FL, USA, 2017; pp. 251–253. [CrossRef]

24. Ahn, D.-G. Direct metal additive manufacturing processes and their sustainable applications for green technology: A review. *Int. J. Precis. Eng. Manuf. Green Technol.* **2016**, *3*, 381–395. [CrossRef]

25. Gu, G.X.; Chen, C.-T.; Richmond, D.J.; Buehler, M.J. Bioinspired hierarchical composite design using machine learning: Simulation, additive manufacturing, and experiment. *Mater. Horiz.* **2018**, *5*, 939–945. [CrossRef]

26. Reddy, N.; Lee, Y.H.; Park, C.H.; Lee, C.S. Prediction of flow stress in Ti–6Al–4V alloy with an equiaxed α+ β microstructure by artificial neural networks. *Mater. Sci. Eng. A* **2008**, *492*, 276–282. [CrossRef]

27. Lubbers, N.; Lookman, T.; Barros, K. Inferring low-dimensional microstructure representations using convolutional neural networks. *Phys. Rev. E* **2017**, *96*, 052111. [CrossRef]

28. Albuquerque, V.H.C.d.; Alexandria, A.R.d.; Cortez, P.C.; Tavares, J.M.R. Evaluation of multilayer perceptron and self-organizing map neural network topologies applied on microstructure segmentation from metallographic images. *NDT E Int.* **2009**, *42*, 644–651. [CrossRef]

29. Jared, B.H.; Aguilo, M.A.; Beghini, L.L.; Boyce, B.L.; Clark, B.W.; Cook, A.; Kaehr, B.J.; Robbins, J. Additive manufacturing: Toward holistic design. *Scr. Mater.* **2017**, *135*, 141–147. [CrossRef]

30. Haase, C.; Tang, F.; Wilms, M.B.; Weisheit, A.; Hallstedt, B. Combining thermodynamic modeling and 3D printing of elemental powder blends for high-throughput investigation of high-entropy alloys–Towards rapid alloy screening and design. *Mater. Sci. Eng. A* **2017**, *688*, 180–189. [CrossRef]

31. Salzbrenner, B.C.; Rodelas, J.M.; Madison, J.D.; Jared, B.H.; Swiler, L.P.; Shen, Y.-L.; Boyce, B.L. High-throughput stochastic tensile performance of additively manufactured stainless steel. *J. Mater. Process. Technol.* **2017**, *241*, 1–12. [CrossRef]

32. Niendorf, T.; Leuders, S.; Riemer, A.; Brenne, F.; Tröster, T.; Richard, H.A.; Schwarze, D. Functionally graded alloys obtained by additive manufacturing. *Adv. Eng. Mater.* **2014**, *16*, 857–861. [CrossRef]

33. Collins, P.; Banerjee, R.; Banerjee, S.; Fraser, H. Laser deposition of compositionally graded titanium–vanadium and titanium–molybdenum alloys. *Mater. Sci. Eng. A* **2003**, *352*, 118–128. [CrossRef]

34. Banerjee, R.; Collins, P.; Bhattacharyya, D.; Banerjee, S.; Fraser, H. Microstructural evolution in laser deposited compositionally graded α/β titanium-vanadium alloys. *Acta Mater.* **2003**, *51*, 3277–3292. [CrossRef]

35. Banerjee, R.; Bhattacharyya, D.; Collins, P.; Viswanathan, G.; Fraser, H. Precipitation of grain boundary α in a laser deposited compositionally graded Ti–8Al–xV alloy–an orientation microscopy study. *Acta Mater.* **2004**, *52*, 377–385. [CrossRef]

36. Joseph, J.; Jarvis, T.; Wu, X.; Stanford, N.; Hodgson, P.; Fabijanic, D.M. Comparative study of the microstructures and mechanical properties of direct laser fabricated and arc-melted Al_xCoCrFeNi high entropy alloys. *Mater. Sci. Eng. A* **2015**, *633*, 184–193. [CrossRef]

37. Gong, X.; Mohan, S.; Mendoza, M.; Gray, A.; Collins, P.; Kalidindi, S. High throughput assays for additively manufactured Ti-Ni alloys based on compositional gradients and spherical indentation. *Integr. Mater. Manuf. Innov.* **2017**, *6*, 218–228. [CrossRef]

38. Khosravani, A.; Cecen, A.; Kalidindi, S.R. Development of high throughput assays for establishing process-structure-property linkages in multiphase polycrystalline metals: Application to dual-phase steels. *Acta Mater.* **2017**, *123*, 55–69. [CrossRef]

39. Weaver, J.S.; Khosravani, A.; Castillo, A.; Kalidindi, S.R. High throughput exploration of process-property linkages in Al-6061 using instrumented spherical microindentation and microstructurally graded samples. *Integr. Mater. Manuf. Innov.* **2016**, *5*, 192–211. [CrossRef]

40. Fernandez-Zelaia, P.; Yabansu, Y.C.; Kalidindi, S.R. A Comparative Study of the Efficacy of Local/Global and Parametric/Nonparametric Machine Learning Methods for Establishing Structure–Property Linkages in High-Contrast 3D Elastic Composites. *Integr. Mater. Manuf. Innov.* **2019**, *8*, 67–81. [CrossRef]

41. Yabansu, Y.C.; Rehn, V.; Hoetzer, J.; Nestler, B.; Kalidindi, S.R. Application of Gaussian process autoregressive models for capturing the time evolution of microstructure statistics from phase-field simulations for sintering of polycrystalline ceramics. *Modell. Simul. Mater. Sci. Eng.* **2019**, *27*, 084006. [CrossRef]

42. Yabansu, Y.C.; Iskakov, A.; Kapustina, A.; Rajagopalan, S.; Kalidindi, S.R. Application of Gaussian process regression models for capturing the evolution of microstructure statistics in aging of nickel-based superalloys. *Acta Mater.* **2019**, *178*, 45–58. [CrossRef]

43. Jung, J.; Yoon, J.I.; Park, H.K.; Kim, J.Y.; Kim, H.S. An efficient machine learning approach to establish structure-property linkages. *Comput. Mater. Sci.* **2019**, *156*, 17–25. [CrossRef]

44. Fernandez-Zelaia, P.; Melkote, S.N. Process-Structure-Property Modeling for Severe Plastic Deformation Processes Using Orientation Imaging Microscopy and Data-Driven Techniques. *Integr. Mater. Manuf. Innov.* **2019**, *8*, 17–36. [CrossRef]

45. Xiong, Y.; Duong, P.L.T.; Wang, D.; Park, S.-I.; Ge, Q.; Raghavan, N.; Rosen, D.W. Data-Driven Design Space Exploration and Exploitation for Design for Additive Manufacturing. *J. Mech. Des.* **2019**, *141*, 101101. [CrossRef]

46. Parvinian, S.; Yabansu, Y.C.; Khosravani, A.; Garmestani, H.; Kalidindi, S.R. High-Throughput Exploration of the Process Space in 18% Ni (350) Maraging Steels via Spherical Indentation Stress–Strain Protocols and Gaussian Process Models. *Integr. Mater. Manuf. Innov.* **2020**, *9*. [CrossRef]

47. Bilionis, I.; Zabaras, N.; Konomi, B.A.; Lin, G. Multi-output separable Gaussian process: Towards an efficient, fully Bayesian paradigm for uncertainty quantification. *J. Comput. Phys.* **2013**, *241*, 212–239. [CrossRef]

48. Hu, Z.; Mahadevan, S. Uncertainty quantification and management in additive manufacturing: Current status, needs, and opportunities. *Int. J. Adv. Manuf. Technol.* **2017**, *93*, 2855–2874. [CrossRef]

49. Chi, G.; Hu, S.; Yang, Y.; Chen, T. Response surface methodology with prediction uncertainty: A multi-objective optimisation approach. *Chem. Eng. Res. Des.* **2012**, *90*, 1235–1244. [CrossRef]

50. Talapatra, A.; Boluki, S.; Duong, T.; Qian, X.; Dougherty, E.; Arróyave, R. Autonomous efficient experiment design for materials discovery with Bayesian model averaging. *Phys. Rev. Mater.* **2018**, *2*, 113803. [CrossRef]

51. Castillo, A.R.; Joseph, V.R.; Kalidindi, S.R. Bayesian Sequential Design of Experiments for Extraction of Single-Crystal Material Properties from Spherical Indentation Measurements on Polycrystalline Samples. *JOM* **2019**, *71*, 2671–2679. [CrossRef]

52. Griffith, M.L.; Harwell, L.D.; Romero, J.T.; Schlienger, E.; Atwood, C.L.; Smugeresky, J.E. Multi-material processing by LENS. In Proceedings of the 1997 International Solid Freeform Fabrication Symposium, Austin, TX, USA, 11–13 August 1997.

53. Griffith, M.; Keicher, D.; Atwood, C.; Romero, J.; Smugeresky, J.; Harwell, L.; Greene, D. Free form fabrication of metallic components using laser engineered net shaping (LENS). In Proceedings of the 1996 International Solid Freeform Fabrication Symposium, Austin, TX, USA, 12–14 August 1996.

54. Mudge, R.P.; Wald, N.R. Laser engineered net shaping advances additive manufacturing and repair. *Weld. J.* **2007**, *86*, 44.

55. Avallone, E.A. *Marks' Standard Handbook for Mechanical Engineers*; The McGraw-Hill Companies, Inc.: New York, NY, USA, 2007.

56. Smugeresky, J.; Keicher, D.; Romero, J.; Griffith, M.; Harwell, L. Laser engineered net shaping(LENS) process: Optimization of surface finish and microstructural properties. *Adv. Powder Metall. Part Mater.* **1997**, *3*, 21. [CrossRef]

57. Palčič, I.; Balažic, M.; Milfelner, M.; Buchmeister, B. Potential of laser engineered net shaping (LENS) technology. *Mater. Manuf. Process.* **2009**, *24*, 750–753. [CrossRef]

58. Attar, H.; Ehtemam-Haghighi, S.; Kent, D.; Wu, X.; Dargusch, M.S. Comparative study of commercially pure titanium produced by laser engineered net shaping, selective laser melting and casting processes. *Mater. Sci. Eng. A* **2017**, *705*, 385–393. [CrossRef]

59. Bandyopadhyay, A.; Krishna, B.V.; Xue, W.; Bose, S. Application of laser engineered net shaping (LENS) to manufacture porous and functionally graded structures for load bearing implants. *J. Mater. Sci. Mater. Med.* **2009**, *20*, 29. [CrossRef] [PubMed]

60. Jeon, T.; Hwang, T.; Yun, H.; VanTyne, C.; Moon, Y. Control of Porosity in Parts Produced by a Direct Laser Melting Process. *Appl. Sci.* **2018**, *8*, 2573. [CrossRef]

61. Wang, L.; Pratt, P.; Felicelli, S.D.; Kadiri, H.E.; Berry, J.T.; Wang, P.T.; Horstemeyer, M.F. Experimental analysis of porosity formation in laser-assisted powder deposition process. *Miner. Met. Mater. Soc.* **2009**, *1*, 389–396.

62. Li, L. Repair of directionally solidified superalloy GTD-111 by laser-engineered net shaping. *J. Mater. Sci.* **2006**, *41*, 7886–7893. [CrossRef]

63. Korinko, P.; Adams, T.; Malene, S.; Gill, D.; Smugeresky, J. Laser Engineered Net Shaping® for Repair and Hydrogen Compatibility. *Weld. J.* **2011**, *90*, 171–181.

64. Liu, Z.; Cong, W.; Kim, H.; Ning, F.; Jiang, Q.; Li, T.; Zhang, H.-C.; Zhou, Y. Feasibility Exploration of Superalloys for AISI 4140 Steel Repairing using Laser Engineered Net Shaping. *Procedia Manuf.* **2017**, *10*, 912–922. [CrossRef]

65. Wilson, J.M.; Piya, C.; Shin, Y.C.; Zhao, F.; Ramani, K. Remanufacturing of turbine blades by laser direct deposition with its energy and environmental impact analysis. *J. Clean. Prod.* **2014**, *80*, 170–178. [CrossRef]

66. Liu, D.; Lippold, J.C.; Li, J.; Rohklin, S.R.; Vollbrecht, J.; Grylls, R. Laser engineered net shape (LENS) technology for the repair of Ni-base superalloy turbine components. *Metall. Mater. Trans. A* **2014**, *45*, 4454–4469. [CrossRef]

67. Balla, V.K.; Bandyopadhyay, P.P.; Bose, S.; Bandyopadhyay, A. Compositionally graded yttria-stabilized zirconia coating on stainless steel using laser engineered net shaping (LENS™). *Scr. Mater.* **2007**, *57*, 861–864. [CrossRef]

68. Balla, V.K.; DeVasConCellos, P.D.; Xue, W.; Bose, S.; Bandyopadhyay, A. Fabrication of compositionally and structurally graded Ti–TiO$_2$ structures using laser engineered net shaping (LENS). *Acta Biomater.* **2009**, *5*, 1831–1837. [CrossRef]

69. Polanski, M.; Kwiatkowska, M.; Kunce, I.; Bystrzycki, J. Combinatorial synthesis of alloy libraries with a progressive composition gradient using laser engineered net shaping (LENS): Hydrogen storage alloys. *Int. J. Hydrogen Energy* **2013**, *38*, 12159–12171. [CrossRef]

70. Bandyopadhyay, P.P.; Balla, V.K.; Bose, S.; Bandyopadhyay, A. Compositionally graded aluminum oxide coatings on stainless steel using laser processing. *J. Am. Ceram. Soc.* **2007**, *90*, 1989–1991. [CrossRef]

71. Zhang, Y.; Bandyopadhyay, A. Direct fabrication of compositionally graded Ti-Al$_2$O$_3$ multi-material structures using Laser Engineered Net Shaping. *Addit. Manuf.* **2018**, *21*, 104–111. [CrossRef]

72. Hofmann, D.C.; Roberts, S.; Otis, R.; Kolodziejska, J.; Dillon, R.P.; Suh, J.-O.; Shapiro, A.A.; Liu, Z.-K.; Borgonia, J.-P. Developing gradient metal alloys through radial deposition additive manufacturing. *Sci. Rep.* **2014**, *4*, 5357. [CrossRef]

73. Lütjering, G.; Williams, J.C. *Titanium*; Springer-Science & Business Media: Berlin/Heidelberg, Germany, 2007.

74. Bobet, J.L.; Darriet, B. Relationship between hydrogen sorption properties and crystallography for TiMn$_2$ based alloys. *Int. J. Hydrog. Energy* **2000**, *25*, 767–772. [CrossRef]

75. Zhang, F.; Weidmann, A.; Nebe, J.B.; Beck, U.; Burkel, E. Preparation, microstructures, mechanical properties, and cytocompatibility of TiMn alloys for biomedical applications. *J. Biomed. Mater. Res. Part B Appl. Biomater.* **2010**, *94B*, 406–413. [CrossRef]

76. Takada, Y.; Nakajima, H.; Okuno, O.; Okabe, T. Microstructure and Corrosion Behavior of Binary Titanium Alloys with Beta-stabilizing Elements. *Dent. Mater. J.* **2001**, *20*, 34–52. [CrossRef] [PubMed]

77. Murray, J.L. The Mn-Ti (Manganese-Titanium) system. *Bull. Alloy Phase Diagr.* **1981**, *2*, 334–343. [CrossRef]

78. Mitchell, A.; Kawakami, A.; Cockcroft, S. Beta fleck and segregation in titanium alloy ingots. *High Temp. Mater. Process.* **2006**, *25*, 337–349. [CrossRef]

79. Seagle, S.; Yu, K.; Giangiordano, S. Considerations in processing titanium. *Mater. Sci. Eng. A* **1999**, *263*, 237–242. [CrossRef]

80. Bomberger, H.; Froes, F. The melting of titanium. *JOM* **1984**, *36*, 39–47. [CrossRef]

81. Gammon, L.M.; Briggs, R.D.; Packard, J.M.; Batson, K.W.; Boyer, R.; Domby, C.W. Metallography and microstructures of titanium and its alloys. *ASM Handb.* **2004**, *9*, 899–917. [CrossRef]

82. Weaver, J.S.; Priddy, M.W.; McDowell, D.L.; Kalidindi, S.R. On capturing the grain-scale elastic and plastic anisotropy of alpha-Ti with spherical nanoindentation and electron back-scattered diffraction. *Acta Mater.* **2016**, *117*, 23–34. [CrossRef]

83. Vachhani, S.J.; Doherty, R.D.; Kalidindi, S.R. Effect of the continuous stiffness measurement on the mechanical properties extracted using spherical nanoindentation. *Acta Mater.* **2013**, *61*, 3744–3751. [CrossRef]

84. Otsu, N. A threshold selection method from gray-level histograms. *IEEE Trans. Syst. Man Cybern. Part A* **1979**, *9*, 62–66. [CrossRef]

85. Sezgin, M.; Sankur, B. Survey over image thresholding techniques and quantitative performance evaluation. *J. Electron. Imaging* **2004**, *13*, 146–166. [CrossRef]

86. Siddique, M.A.B.; Arif, R.B.; Khan, M.M.R. Digital Image Segmentation in Matlab: A Brief Study on OTSU's Image Thresholding. In Proceedings of the 2018 International Conference on Innovation in Engineering and Technology (ICIET), Dhaka, Bangladesh, 27–28 December 2018; pp. 1–5. [CrossRef]

87. Turner, D.M.; Niezgoda, S.R.; Kalidindi, S.R. Efficient computation of the angularly resolved chord length distributions and lineal path functions in large microstructure datasets. *Modell. Simul. Mater. Sci. Eng.* **2016**, *24*, 075002. [CrossRef]

88. Torquato, S. *Random Heterogeneous Materials: Microstructure and Macroscopic Properties*; Springer Science & Business Media: New York, NY, USA, 2013; Volume 16.

89. Latypov, M.I.; Kühbach, M.; Beyerlein, I.J.; Stinville, J.-C.; Toth, L.S.; Pollock, T.M.; Kalidindi, S.R. Application of chord length distributions and principal component analysis for quantification and representation of diverse polycrystalline microstructures. *Mater. Charact.* **2018**, *145*, 671–685. [CrossRef]

90. Pathak, S.; Stojakovic, D.; Doherty, R.; Kalidindi, S.R. Importance of surface preparation on the nano-indentation stress-strain curves measured in metals. *J. Mater. Res.* **2009**, *24*, 1142–1155. [CrossRef]

91. Pathak, S.; Kalidindi, S.R. Spherical nanoindentation stress–strain curves. *Mater. Sci. Eng. R Rep.* **2015**, *91*, 1–36. [CrossRef]

92. Kalidindi, S.R.; Pathak, S. Determination of the effective zero-point and the extraction of spherical nanoindentation stress–strain curves. *Acta Mater.* **2008**, *56*, 3523–3532. [CrossRef]

93. Hertz, H.; Jones, D.E.; Schott, G.A. *Miscellaneous Papers*; Macmillan and Company: London, UK, 1896.

94. Johnson, K.L.; Johnson, K.L. *Contact Mechanics*; Cambridge University Press: Cambridge, UK, 1987.

95. Oliver, W.C.; Pharr, G.M. An improved technique for determining hardness and elastic modulus using load and displacement sensing indentation experiments. *J. Mater. Res.* **1992**, *7*, 1564–1583. [CrossRef]

96. Li, X.; Bhushan, B. A review of nanoindentation continuous stiffness measurement technique and its applications. *Mater. Charact.* **2002**, *48*, 11–36. [CrossRef]

97. Pathak, S.; Kalidindi, S.R.; Mara, N.A. Investigations of orientation and length scale effects on micromechanical responses in polycrystalline zirconium using spherical nanoindentation. *Scr. Mater.* **2016**, *113*, 241–245. [CrossRef]

98. Pathak, S.; Stojakovic, D.; Kalidindi, S.R. Measurement of the local mechanical properties in polycrystalline samples using spherical nanoindentation and orientation imaging microscopy. *Acta Mater.* **2009**, *57*, 3020–3028. [CrossRef]

99. Patel, D.K.; Al-Harbi, H.F.; Kalidindi, S.R. Extracting single-crystal elastic constants from polycrystalline samples using spherical nanoindentation and orientation measurements. *Acta Mater.* **2014**, *79*, 108–116. [CrossRef]

100. Williams, C.K.; Rasmussen, C.E. *Gaussian Processes for Machine Learning*; MIT Press: Cambridge, MA, USA, 2006; Volume 2.

101. Bishop, C.M. *Pattern Recognition and Machine Learning (Information Science and Statistics)*; Springer: Berlin/Heidelberg, Germany, 2006.

102. Yabansu, Y.C.; Altschuh, P.; Hötzer, J.; Selzer, M.; Nestler, B.; Kalidindi, S.R. A digital workflow for learning the reduced-order structure-property linkages for permeability of porous membranes. *Acta Mater.* **2020**, *195*, 668–680. [CrossRef]

103. Witte, R.S.; Witte, J.S. *Statistics*; Wiley: Hoboken, NJ, USA, 2009.

104. Dette, H.; Munk, A. Testing heteroscedasticity in nonparametric regression. *J. R. Stat. Soc. Ser. B Stat. Methodol.* **1998**, *60*, 693–708. [CrossRef]

105. Weaver, J.S.; Kalidindi, S.R. Mechanical characterization of Ti-6Al-4V titanium alloy at multiple length scales using spherical indentation stress-strain measurements. *Mater. Des.* **2016**, *111*, 463–472. [CrossRef]

106. Mohammed, M.T.; Khan, Z.A.; Siddiquee, A.N. Beta titanium alloys: The lowest elastic modulus for biomedical applications: A review. *Int. J. Chem. Mol. Nucl. Mater. Metall. Eng.* **2014**, *8*, 726–731. [CrossRef]

107. Laheurte, P.; Prima, F.; Eberhardt, A.; Gloriant, T.; Wary, M.; Patoor, E. Mechanical properties of low modulus β titanium alloys designed from the electronic approach. *J. Mech. Behav. Biomed. Mater.* **2010**, *3*, 565–573. [CrossRef] [PubMed]

108. Santos, P.F.; Niinomi, M.; Liu, H.; Cho, K.; Nakai, M.; Itoh, Y.; Narushima, T.; Ikeda, M. Fabrication of low-cost beta-type Ti–Mn alloys for biomedical applications by metal injection molding process and their mechanical properties. *J. Mech. Behav. Biomed. Mater.* **2016**, *59*, 497–507. [CrossRef] [PubMed]

109. Welsch, G.; Boyer, R.; Collings, E. *Materials Properties Handbook: Titanium Alloys*; ASM International: Materials Park, OH, USA, 1993.

110. Patel, D.K.; Kalidindi, S.R. Correlation of spherical nanoindentation stress-strain curves to simple compression stress-strain curves for elastic-plastic isotropic materials using finite element models. *Acta Mater.* **2016**, *112*, 295–302. [CrossRef]

111. Cho, K.; Niinomi, M.; Nakai, M.; Liu, H.; Santos, P.F.; Itoh, Y.; Ikeda, M.; Narushima, T. Improvement in mechanical strength of low-cost β-type Ti–Mn alloys fabricated by metal injection molding through cold rolling. *J. Alloys Compd.* **2016**, *664*, 272–283. [CrossRef]

112. Alshammari, Y.; Yang, F.; Bolzoni, L. Mechanical properties and microstructure of Ti-Mn alloys produced via powder metallurgy for biomedical applications. *J. Mech. Behav. Biomed. Mater.* **2019**, *91*, 391–397. [CrossRef]

113. Santos, P.F.; Niinomi, M.; Cho, K.; Nakai, M.; Liu, H.; Ohtsu, N.; Hirano, M.; Ikeda, M.; Narushima, T. Microstructures, mechanical properties and cytotoxicity of low cost beta Ti–Mn alloys for biomedical applications. *Acta Biomater.* **2015**, *26*, 366–376. [CrossRef]

Publisher's Note: MDPI stays neutral with regard to jurisdictional claims in published maps and institutional affiliations.

Article

A Strategy for Dimensionality Reduction and Data Analysis Applied to Microstructure–Property Relationships of Nanoporous Metals

Norbert Huber [1,2]

[1] Institute of Materials Mechanics, Helmholtz-Zentrum Hereon, Max-Planck-Str. 1, 21502 Geesthacht, Germany; norbert.huber@hereon.de; Tel.: +49-4152-87-2501

[2] Institute of Materials Physics and Technology, Hamburg University of Technology, Eißendorfer Str. 42, 21073 Hamburg, Germany

Abstract: Nanoporous metals, with their complex microstructure, represent an ideal candidate for the development of methods that combine physics, data, and machine learning. The preparation of nanporous metals via dealloying allows for tuning of the microstructure and macroscopic mechanical properties within a large design space, dependent on the chosen dealloying conditions. Specifically, it is possible to define the solid fraction, ligament size, and connectivity density within a large range. These microstructural parameters have a large impact on the macroscopic mechanical behavior. This makes this class of materials an ideal science case for the development of strategies for dimensionality reduction, supporting the analysis and visualization of the underlying structure–property relationships. Efficient finite element beam modeling techniques were used to generate ~200 data sets for macroscopic compression and nanoindentation of open pore nanofoams. A strategy consisting of dimensional analysis, principal component analysis, and machine learning allowed for data mining of the microstructure–property relationships. It turned out that the scaling law of the work hardening rate has the same exponent as the Young's modulus. Simple linear relationships are derived for the normalized work hardening rate and hardness. The hardness to yield stress ratio is not limited to 1, as commonly assumed for foams, but spreads over a large range of values from 0.5 to 3.

Keywords: nanoporous metals; open-pore foams; FE-beam model; data mining; mechanical properties; hardness; machine learning; principal component analysis; structure–property relationship; microcompression; nanoindentation

Citation: Huber, N. A Strategy for Dimensionality Reduction and Data Analysis Applied to Microstructure–Property Relationships of Nanoporous Metals. *Materials* **2021**, *14*, 1822. https://doi.org/10.3390/ma14081822

Academic Editors: Francisco Chinesta, Elías Cueto and Benjamin Klusemann

Received: 12 March 2021
Accepted: 5 April 2021
Published: 7 April 2021

Publisher's Note: MDPI stays neutral with regard to jurisdictional claims in published maps and institutional affiliations.

1. Introduction

Nanoporous gold (np-Au) made by dealloying can be produced as macroscopic objects that exhibit a bi-continuous network of nanoscale pores and solid "ligaments" connected in nodes. An overview of the fascinating morphologies and mechanical properties of this material is provided in [1–3]. The skeleton of the structure is formed by ligaments, which can be controlled in their average diameter by altering the dealloying conditions, thus allowing one to examine the impact of the ligament size on the macroscopic mechanical properties [4,5]. It has been recently demonstrated that the dealloying process can be applied sequentially and allows one to produce hierarchically organized nanoporous metals with superior macroscopic properties compared to similar materials with only one hierarchy level [6].

So far, even for one hierarchy level, no model exists that allows for the prediction of the macroscopic mechanical properties based on the parameters used in the sample preparation. Recently, the evolution of the ligament size and the network connectivity during thermal treatment was modeled with kinetic Monte Carlo simulations [7] for a large range of solid fractions, but the connection to the macroscopic mechanical properties is still missing. For a selected microstructure, this is realized by conventional meshing and finite element (FE) simulation, e.g., as shown in [8], but from this point, it is still a long way to go towards an all-inclusive process–microstructure–property model that handles all required steps along a fully automated work flow and at the required efficiency. An overview of the elements needed for such a work flow based on efficient simulation models, data mining, and AI is presented in [9].

A key element represents the relationship that efficiently translates a set of microstructural parameters and material properties of the solid phase into macroscopic properties. Together with the structural information from, e.g., high-resolution 3D tomography and image analysis [10,11], all relevant aspects are currently under development. As pointed out in [9], they altogether will allow for an efficient scan of large multidimensional parameter spaces of descriptors and reliably predict the macroscopic mechanical properties for any assumed constitutive law on the level of a single ligament. Moving from scarce data to rich data allows for data mining of the fundamental structure–property relationships. The objective is to derive robust approximations that generalize the available data and support our understanding of the underlying physics well beyond the application of machine learning as a black box method.

In this work, we concentrate on the relationship that allows predicting mechanical properties based on microstructural information or, formulated as an inverse problem, enables us to determine microstructural descriptors from macroscopic test data. Due to the complex morphology of this material, FE modeling of np-Au with all its structural complexity is highly challenging. Two general paths exist, which are summarized in [9]. One route uses random structures (spinodal decomposition, leveled waves); the second is based on unit cells (Gibson–Ashby, gyroid, diamond). The mechanical behavior of random structures is usually predicted with molecular dynamics (MD) simulations [12–14] or with continuum mechanics using FE-solid or voxel models [8,15,16]. In combination with plasticity, also the FE-models lead to large computing times and allow only for a very limited number of simulations. Furthermore, the limited model size makes it extremely difficult to simulate a nanoindentation test that averages over sufficient features, such that it can be analyzed like an experiment. One of the rare examples that goes in this direction is the work of Farkas et al. [14], which presents a MD simulation of nanoindentation in a single crystal with a relative density of 0.67 and ligament diameter of 2 nm.

As pointed out in [9], FE-beam models provide the efficiency and flexibility needed for the generation of larger data sets and, at the same time, allow for an independent variation of all structural and material parameters of interest. This modeling technique has been successfully applied in studying the mechanical behavior of foams [17–22] and nanoporous metals [10,23–27]. Research in this field concentrated mainly on the anisotropy of the macroscopic elastic properties as well as aspects of the structure–property relationships for elastic–plastic macroscopic compression. Until recently, the quantitative correct prediction of materials with relative densities >10% was limited to cylindrical ligament shapes [26]. The nodal correction proposed by Odermatt et al. enables us to expand FE simulations towards variations of the ligament shape from concave to convex [27]. The advantage of this approach is that the computational efficiency of FE-beam models is maintained. This paves the way for handling hundreds of simulations with a predictive model that is at the same time large enough for the simulation of nanoindentation.

The scope of this work is to study the influence of microstructure and material parameters on the macroscopic response of a porous metal. We will investigate the macroscopic behavior under compression as well as nanoindentation. For scanning the multidimensional parameter space, a highly efficient simulation model is required. Furthermore, the model set-up should allow for the independent variation of all important structural inputs. To this end, we use a representative volume element (RVE) that approximates the complex morphology of an open pore material by a diamond structure [23,24]. Using this unit cell, it is possible to define the degree of randomization and connectivity of the structure [28]. Together with the material parameters defining the mechanical behavior of the solid phase, this generates a highly dimensional parameter space that is hard to scan in a dense manner by numerical simulations. The dimensionality of the problem and limited number of simulations makes analysis of the underlying structure–property relationships very challenging. If we limit the number of parameters to five (two material and three microstructural parameters) and the number of variations per parameter to three, a systematic variation with one parameter at a time would end up with 243 simulations, which is already at the limit of the computer's capacity. Adding more parameters or increasing the number of increments is almost impossible. Therefore, this investigation requires a strategy that exploits all available methods that contribute to reduce the dimensionality of the problem.

In this sense, the present work also serves as a guide, demonstrating how such a problem can be tackled systematically by means of a dimensional analysis, inclusion of a priori knowledge about the physical problem at hand, data generation strategies, principal component analysis, machine learning, and visualization. Along this path, Section 2 describes the FE-beam models used for generation of the data for macroscopic compression and nanoindentation. Sections 3 and 4 deal with dimensionality reduction of the macroscopic compression and nanoindentation problem, respectively, where both sections follow the same methodology. Finally, it is shown that for important dependencies, simple mathematical formulations can be derived that relate the major influences of microstructure and mechanical properties to the macroscopic response.

2. FE-Model and Data Generation

FE-beam modeling is used to predict the macroscopic response of nanoporous metals during macroscopic compression and nanoindentation. The generation of the representative volume element (RVE) is established in the literature and is described only briefly in Section 2.1. In contrast, the simulation of nanoindentation is novel. The incorporation of a conical indenter and strategies for achieving an efficient simulation model that copes with the nonlinearities arising from the contact problem is described in Section 2.2.

2.1. Macroscopic Compression

The FE software Abaqus was used for the numerical simulation of the RVE [29]. The model generation for macroscopic compression followed [23,24,27,28] and was organized hierarchically along the workflow presented in Figure 1. This workflow was programmed object oriented in Python with classes for the different hierarchy levels, allowing for scripting of the RVE generation and job submission within loops for the variation of input parameters. A postprocessing script handled the simulation analysis and database generation.

The model generation started at the ligament level, where the ligament axis is discretized in N_{elem} FE beam elements with circular cross-section and variable radius r. The ligament shape is defined along the axis according to [10,11] by $r^*_{sym} = r_{mid}/r_{end}$ and r_{end}/l, where r_{mid} and r_{end} denote the ligament radius in the middle and at the ends, respectively, and l is the ligament length in a diamond unit cell. Together with the topology, the set of ligament geometry parameters r_{mid}, r_{end}, and l define the solid fraction φ_0 before randomization.

Odermatt et al. [27] developed nodal corrections for 16 ligament shapes that allow for a quantitative prediction of the elastic–plastic response of the RVE up to macroscopic strains of 20%. Details about the ligament geometries, initial solid fractions, and the nodal correction approach can be found in [27]. The extension of the nodal corrected zones is visible in the second column of Figure 1, where nodal corrected elements in the diamond unit cell are displayed in orange. With the nodal correction set "on", their material parameters were modified such that the deformation behavior of the unit cell corresponded to that of an FE solid model of the same ligament shape. Preliminary simulations for decreasing number of elements using a unit cell with periodic boundary conditions confirmed that the nodal correction by [27] performed well in the range from 20 down to 6 FE elements per ligament for both geometries listed in Table 1. Therefore,

6 elements per ligament were chosen in this work for which the relative error in macroscopic stiffness and strength was within 15% error relative to the results of the FE solid model.

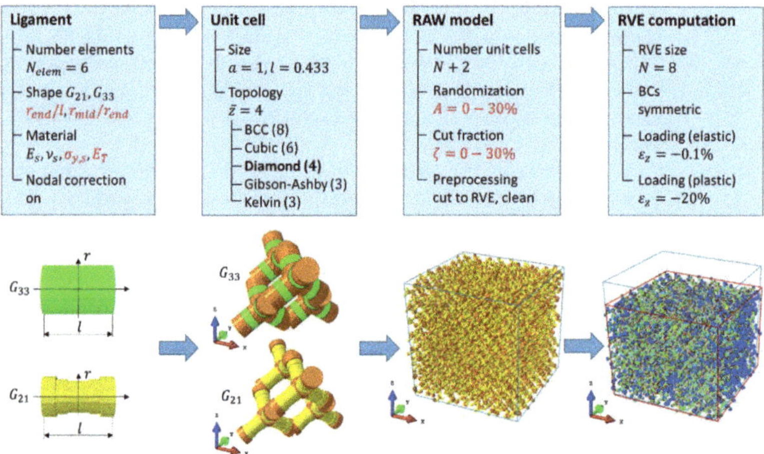

Figure 1. Workflow for generating a representative volume element (RVE) in four steps: (i) ligament, (ii) unit cell, (iii) RAW model after cutting and cleaning, and (iv) computation of the RVE. Varied parameters are highlighted in red color.

Table 1. Ligament shapes used for the generation of two data sets in the low and a high solid fraction regime, respectively.

Shape	r^*_{sym}	r_{end}/l	φ_0
G_{21}	0.5	0.289	0.1232
G_{33}	1.0	0.346	0.3574

For generating an RVE of size N, the unit cell is copied $N + 2$ times (origin at $[-1, -1, -1]$) in each coordinate direction, followed by the randomization of the structure. The degree of randomization is defined by the parameter A, which corresponds to a random displacement in space applied to the connecting nodes by an amplitude A, which is given as a fraction of the unit cell size a [23,24]. Alternatively, one can also choose to displace an FE node in the mid-section of the ligaments by this magnitude normal to the ligament axis [27]. The randomization can be calibrated via the elastic Poisson's ratio and is typically $A = 0.23$ [24].

Because the coordination of the diamond structure of $\bar{z} = 4$ is too high in comparison to experimental observations [10,30,31], the connectivity can be reduced by random cutting of a fraction ζ of the ligaments [28]. For diamond, the percolation threshold is reached for sufficiently large RVEs at a cut fraction of $\zeta \to 0.5$, where the average coordination number approaches $\bar{z} \to 2$. For models of smaller size the percolation threshold is reached at lower values and is sensitive to the random realization. In combination with randomly cut ligaments, the randomization A can be reduced to values close to 0 to reach the elastic Poisson's ratio measured in experiments [28]. Therefore, we chose these two parameters independent of each other and within comparably large ranges of $0 \leq A \leq 0.3$ and $0 \leq \zeta \leq 0.3$.

The resulting RAW model of size $10 \times 10 \times 10$ unit cells is randomly distorted and can contain free floating ligaments due to random cuts. A cleaned RVE is generated by cutting the RAW model to a cubic volume of size $N = 8$ (origin at $[0, 0, 0]$) by removing all elements outside of this volume. Free floating ligaments are removed by two subsequent cleaning cycles that eliminate dangling ligaments and then re-attach element by element those ligaments that are connected to the residual core of the ligament network. For more details, the reader is referred to the supplementary material that is provided in [28]. The result of the preprocessing is an RVE of dimensions $8 \times 8 \times 8$ unit cells with plane boundaries, consisting of 512 diamond unit cells with a total of 8192 ligaments and 49,152 FE-elements ($A = 0$, $\zeta = 0$). Symmetry boundary conditions are applied to FE-nodes in the planes $x = 0$, $y = 0$, and $z = 0$, while macroscopic compression is applied at the top face at the position $z = N$.

For simplicity, the model was generated such that the unit cell size corresponds to a unit size of 1 mm. Realistic microstructural dimensions of the ligament and the pore size can be achieved by self-similar scaling of the model to a desired characteristic size, e.g., a ligament diameter of 20–150 nm [5]. Because the material law does not account for size effects, the resulting macroscopic behavior is not affected by such a scaling. However, when the effect of the surface energy is included, the ligament size is important; then also the applied electrode potential must be defined [32]. These two parameters allow for switching of the strength and the plastic Poisson's ratio during macroscopic deformation of the material.

A and ζ are dimensionless structural parameters describing the random distortion of the connecting nodes as fraction of the unit cell size and the fraction of randomly cut ligaments, respectively. Both parameters modify the solid fraction relative to the initial solid fraction φ_0. According to Roschning et al. [24], we should account for the distortion of the ligament axis by A by an increase in solid fraction by using

$$\frac{\varphi_A}{\varphi_0} = 1 + 0.15A + 2.91A^2, \tag{1}$$

whereas the random cutting ζ removes a fraction of ligaments and, therefore, mass from the model [28]

$$\frac{\varphi_\zeta}{\varphi_0} = 1 - \zeta. \tag{2}$$

If the RVE is large enough, Equations (1) and (2) can be combined as

$$\frac{\varphi}{\varphi_0} = (1 - \zeta)\left(1 + 0.15A + 2.91A^2\right). \tag{3}$$

It should be noted that the random cutting ζ can lead to a mechanical deactivation of whole regions that are still part of the model. Therefore, φ_ζ should not be interpreted as effective solid fraction φ_{eff} that represents the load bearing mass [33].

In view of the number of parameters that may play a role, we limited the structural variation to the randomization A and the cut fraction ζ and kept all other structural parameters within each data set constant (ligament aspect ratio r_{end}/l, ligament shape r_{sym}^*). Two data sets for ligament shapes G_{21} and G_{33} (see Table 1) were created, covering a large range from very low ($\varphi_0 \sim 12\%$) to very high ($\varphi_0 \sim 36\%$) solid fractions. Because the porosity was computed from $1 - \varphi_0$, the porosity ranged from $\sim 64\%$ to $\sim 88\%$.

We used nanoporous gold (np-Au) as model material, because in terms of microstructure and mechanical properties this is the best investigated material of a variety of nanoporous metals reported in the literature. The chosen material behavior is plasticity with linear isotropic hardening [23]. This adds two material parameters denoted as yield stress $\sigma_{y,s}$ and work hardening rate $E_{T,s}$; the subscript s denotes that both parameters are a property of the solid phase, which makes up the 3D network. Both depend on the ligament diameter, which can be manipulated during the sample preparation of the material via the Au/Ag ratio, dealloying conditions, and heat treatment, as demonstrated in [5,32]. The elastic constants for gold are known and were kept constant for all simulations: Young's modulus $E_s = 80$ GPa, Poisson's ratio $\nu_s = 0.42$. An example of a deformed RVE ($A = 18\%$, $\zeta = 26\%$) is shown in Figure 2a. The stress is evenly distributed over the length of the RVE, which indicates that the overall deformation is homogeneous despite the local structural variations due to the randomization of the ligament network.

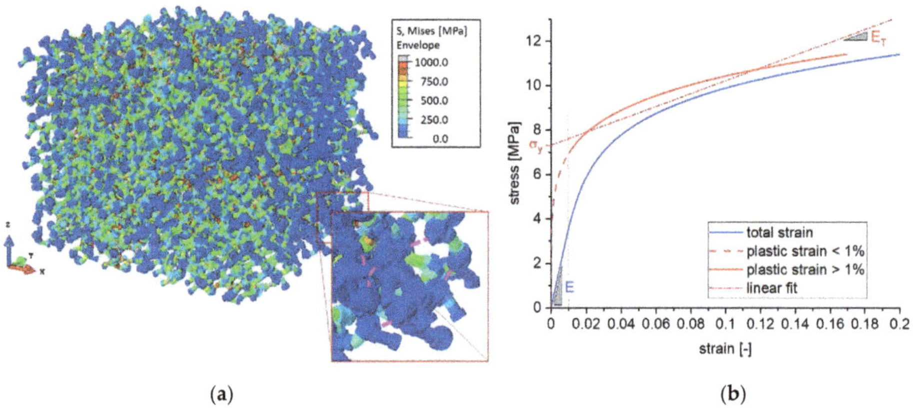

Figure 2. (**a**) RVE consisting of $8 \times 8 \times 8$ unit cells with $A = 18\%$ and $\zeta = 26\%$ after compression with 20% strain in the negative z-direction. The purple dashed curves in the magnified image shown on the right side indicate the axis of some cut ligaments. Due to the missing load transmission, the remaining dangling parts show zero stress (blue color). (**b**) Simulation output in the form of stress–strain and stress–plastic strain curves from which the macroscopic yield stress and work hardening rate are determined.

This model makes up a set of variable inputs consisting of 5 independent parameters:

$$X = \left(\varphi_0, A, \zeta, \sigma_{y,s}, E_{T,s} \right). \tag{4}$$

For each initial solid fraction, the remaining parameters are randomly set for each simulation within the ranges $0 \leq A \leq 0.3$, $0 \leq \zeta \leq 0.3$, $20 \text{ MPa} \leq \sigma_{y,s} \leq 1000 \text{ MPa}$, and $1 \text{ GPa} \leq E_{T,s} \leq 10 \text{ GPa}$, which cover the known range of experimental data. The random distribution of the parameters is uniform for A, ζ, $\log \sigma_{y,s}$, and $\log E_{T,s}$. Each parameter set is stored together with the job number, which uniquely connects microscopic to macroscopic compression as well as nanoindentation properties in the data processing in Sections 3 and 4. The random choice of the parameter sets has the advantage that the parameter space is evenly filled while no parameter is computed more than once. This avoids patterns that might be unwantedly recognized by the machine learning algorithms. Furthermore, the parameter space can continued to be filled if it turns out that the number of patterns is not sufficient for the analysis. This is particularly useful when the simulations are computationally expensive. For an example where this strategy is applied in combination with artificial neural networks for solving a complex inverse problem in nanoindentation, the reader is referred to [34].

The resulting compression behavior of each pattern is represented by 5 dependent properties:

$$Y = \left(E, \nu, \nu_p, \sigma_y, E_T \right), \tag{5}$$

where E, ν, σ_y, and E_T denote the macroscopic Young's modulus, elastic Poisson's ratio, yield stress, and work hardening rate, respectively. The computation of the plastic Poisson's ratio ν_p follows [32]

$$\nu_p = -\frac{\delta \varepsilon_\perp}{\delta \varepsilon_\parallel}, \tag{6}$$

where $\delta \varepsilon_\perp$ and $\delta \varepsilon_\parallel$ are increments of true strain normal and parallel to the loading direction, respectively. Because ν_p changes during plastic compression, it is measured at 10% plastic compression strain. As demonstrated in Figure 2b, the predicted stress–plastic strain data is linearly fitted for plastic strains $> 1\%$ for obtaining the macroscopic yield stress σ_y and work hardening rate E_T.

2.2. Nanoindentation

For the simulation of nanoindentation, the model described in Section 2.1 is extended by adding a conical indenter with an angle of 140.6°. For this angle, the volume-to-depth ratio of the conical indenter corresponds to that of a Berkovich tip. Details on the simulation of nanoindentation for solids and thin films can be found, e.g., in [34]. Due to the numerous ligaments that get in contact during the indentation process, an explicit dynamic analysis was required for achieving convergence. A robust load signal was produced by attaching dashpots at the free boundaries (see Figure 3) to damp elastic waves induced by the multiple contact events during the dynamic indentation process.

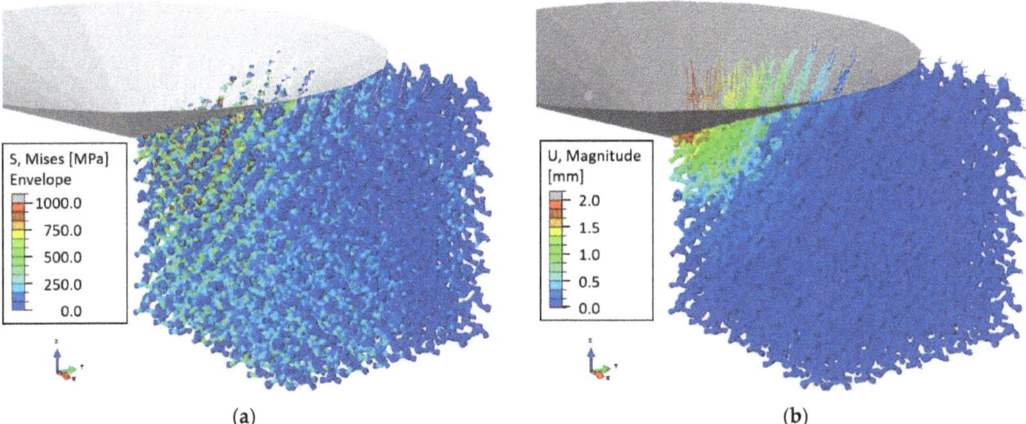

Figure 3. Nanoindentation model with a conical indenter displaced by 2 unit cells at a speed of 20 mm/s. Dashpots are attached at FE nodes located at free boundaries to stabilize oscillations in the dynamic simulation. Images show the deformation of the RVE at the end of the loading phase, where the color corresponds to (**a**) von Mises stress, and (**b**) displacement magnitude.

It can be seen from Figure 3 that the contact of the indenter, modeled as a rigid body, is established with the axis of the beam elements. Therefore, the upper half of the ligaments in contact peek out on the upper side of the indenter surface. Contact among the ligaments is not considered. In principle, this is possible in Abaqus Explicit, but the contact is limited to a pair of a rendered element surface and the axis of a second element. Preliminary studies with this indentation model revealed that such events happen rarely and at a very late stage of the indentation and, therefore, can be neglected in the total force on the indenter. It should be noted that this situation can change once we work with real microstructures and with a contact formulation that accounts for the surface of both contacting ligaments.

The calibration of the indenter velocity and the dashpot parameter is presented in Figure 4 for ligament geometry G_{21} ($\varphi_0 = 0.12$) with $\sigma_{y,s} = 200$ MPa, $E_{T,s} = 6$ GPa, and a randomization $A = 0.23$ [32]. For simplicity, effects of the surface energy are not included, and the cut fraction is set to $\zeta = 0$. For uniaxial compression, the predicted stress–strain curve yields the following macroscopic mechanical properties: $E = 1.9$ GPa, $v = 0.178$, $\sigma_y = 17.8$ MPa, and $E_T = 108$ MPa.

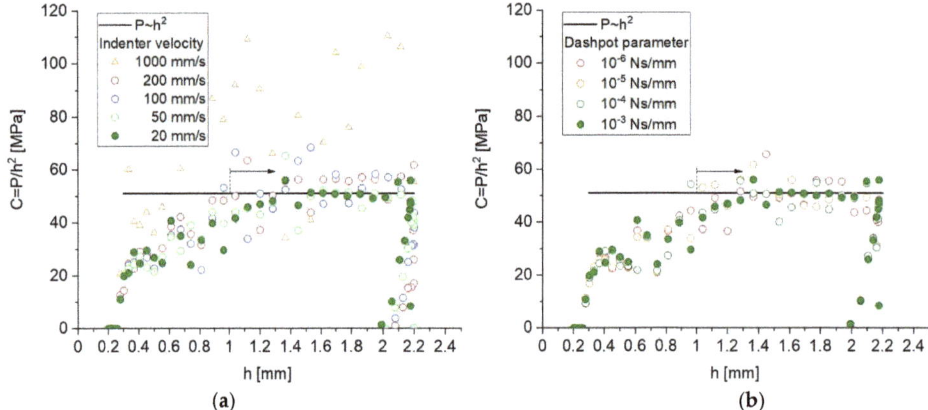

Figure 4. Parametric study for adjustment of (**a**) indenter velocity with smooth step function for a dashpot parameter of 10^{-3} Ns/mm and (**b**) dashpot parameter at fixed indenter velocity of 20 mm/s.

For a conical or pyramidal indenter, the load P always increases with the square of the indentation depth h [34], as soon as the indented material acts like a continuum. Therefore, a plot of P/h^2 vs. h should tend towards a constant value. This is reached for $h > 1$ mm (~1 unit cell), indicating that at this depth value, a sufficient number of ligaments are in contact, and the solution homogenizes over enough microstructural elements. It can be seen from Figure 4 that the results show large scatter for high loading rates and for a low dashpot constant. The load–depth curves converged into a sufficiently steady state solution for an indenter velocity of 20 mm/s, combined with a smooth step function and a dashpot parameter of 10^{-3} Ns/mm. These settings were used for all following simulations, including those shown in Figure 3. The displacement magnitude in Figure 3b shows negligible deformation at the free boundaries, which suggests that the RVE is sufficiently large. With this model and parameter setting, the total CPU time per simulation is ~60 CPUh. Generating a data set with 100 simulations for a selected ligament shape requires ~1 week in real time by parallel computing on 16 CPUs.

For $h > 1$ mm, P_i/h_i^2 values are averaged to compute the leading constant C describing the loading curve $P = Ch^2$. A robust hardness value can be computed by $H = P_t/A_c$, where $P_t = Ch_t^2$ is the load at maximum indentation depth, $A_c = \pi a_c^2$ is the contact area, and a_c is the contact radius at this depth. For the example shown in Figures 3 and 4, we obtain $a_c = 5.3$ mm and a hardness of $H = 2.79$ MPa. Thus, the hardness value is significantly lower than the macroscopic yield stress, which is $\sigma_y = 17.8$ MPa, whereas the common assumption for foams is that $H = \sigma_y$ [35–38]. This motivates a detailed investigation of the dependence of H/σ_y with respect to possible effects caused by the network geometry (randomness, connectivity) and elastic–plastic material properties of the ligaments, which is presented in Section 4. The data generation for the nanoindentation simulations uses the same parameter sets as those used for the simulation of macroscopic compression in Section 3, i.e., for each solid fraction, we performed 100 simulations for macroscopic compression and another 100 simulations for nanoindentation. In a few cases, the simulations of the macroscopic compression did not converge. These parameter sets were removed from both databases to avoid confusion in the analysis that combines macroscopic compression with nanoindentation data.

3. Macroscopic Compression

In the following sections, we reduced the dimensionality of the problem to extract relationships from our data that can be visualized, discussed, and, in the best case, modeled with simple mathematical functions. Our strategy consisted of three steps: (i) dimensional analysis, (ii) principal component analysis, and (iii) visualization and modeling of the relationship with a minimum number of inputs. The dimensional analysis [39] makes use of the physics background and the Buckingham π theorem to reduce the problem without loss of accuracy. This turned out to be a useful approach that should always be placed as a first step of feature engineering, because it ensures that the basic physics is incorporated in the input and output data, while at the same time the machine learning algorithms are relieved and their generalization capability is substantially increased [34,40,41].

Principal component analysis (PCA) was applied in conjunction with a multi-layer perceptron (MLP) algorithm using the scikit-learn package [42]. The MLP, also known as artificial neural networks, allows for the analysis of patterns consisting of multiple inputs and outputs with respect to underlying nonlinear dependencies. For details and applications, the reader is referred to [43–45]. After the dimensionality of the problem was reduced, comparably compact MLPs consisting of two hidden layers with 3 and 2 neurons were used for approximation and visualization of the data. This is possible, when the relationship of interest is sufficiently represented by the selected inputs.

3.1. Dimensional Analysis

The mechanical behavior of the RVE can written in form of dependencies for the elastic and plastic macroscopic properties

$$(E, \nu) = f_e\left(\frac{r_{mid}}{r_{end}}, \frac{r_{end}}{l}, E_s, \nu_s, A, \zeta\right) \tag{7}$$

and

$$(\sigma_y, E_T, \nu_p) = f_p\left(\frac{r_{mid}}{r_{end}}, \frac{r_{end}}{l}, E_s, \nu_s, \sigma_{y,s}, E_{T,s}, A, \zeta\right), \tag{8}$$

respectively. Assuming that the ligament shape is sufficiently represented by the initial solid fraction, Equations (7) and (8) simplify to

$$(E, \nu) = g_e(\varphi_0, E_s, \nu_s, A, \zeta), \tag{9}$$

$$(\sigma_y, E_T, \nu_p) = g_p(\varphi_0, E_s, \nu_s, \sigma_{y,s}, E_{T,s}, A, \zeta). \tag{10}$$

First, we used a priori knowledge in form of the Gibson–Ashby scaling law $E/E_s = C_E \varphi^2$ [35]. The leading constant C_E depends on the unit cell geometry, which in our case was defined by the diamond structure and the chosen ligament shape. To simplify Equation (9) with respect to the Young's modulus, we can assume that the Poisson's ratio of the ligaments has no effect on the macroscopic deformation of the RVE, which results mainly from bending of the ligaments [23]. Combining both aspects and include Equation (3) for computing the solid fraction, we can reduce Equation (9) to a dependence of only two microstructural descriptors,

$$\frac{E}{E_s \varphi^2} = g_E^*(A, \zeta), \tag{11}$$

which can be evaluated easily by visualization of the data in a 3D plot. If such a plot confirms Equation (11), the varying ligament shape is sufficiently represented in the solid fraction φ. Furthermore, g_E^* represents a generalized Gibson–Ashby law that considers the dependence from the degree of randomization and cuts of the 3D network, which is not captured simply by the solid fraction. It also extends the master curve proposed in [28], which was produced using perfectly ordered RVEs, a single solid fraction, and constant material behavior.

Along the same line of thinking, it follows for the simplification of Equation (9) with respect to Poisson's ratio that a dimensionless macroscopic property can only depend on dimensionless microscopic quantities, i.e., the Young's modulus E_s plays no role. In the same way as before, we can remove a dependence of ν_s. The macroscopic Poission's ratio can be understood as the result of the translation of the vertical compression deformation into a lateral expansion by the architecture of the deforming 3D network, defined by A and ζ. This argument is in line with Gibson and Ashby, who stated that the Poisson's ratio is expected to be independent of the relative density [46]. Thus, we get

$$\nu = g_\nu^*(A, \zeta). \tag{12}$$

Concerning the increased number of independent parameters in Equation (10), dimensionality reduction would support both their understanding and modeling of their relationships responsible for the plastic response. Before going into the analysis of the data, it is useful to rewrite this equation in dimensionless form. Again, we can assume that E_s and ν_s have no effect. For plasticity, this can only be assumed as long as $\sigma_{y,s} \ll E_s$ and $E_{T,s} \ll E_s$. Otherwise, we would combine comparable contributions of elastic and plastic deformation in the macroscopic response of the RVE, which requires the consideration of two dimensionless parameters for describing the elastic plastic behavior, namely $\sigma_{y,s}/E_s$ and $E_{T,s}/E_s$. Using the Buckingham π theorem [39], we can eliminate one more argument without loss of generality. One way is to normalize the macroscopic properties on the left side by their respective solid properties in the form

$$\left(\frac{\sigma_y}{\sigma_{y,s}}, \frac{E_T}{E_{T,s}}, \nu_p \right) = \hat{g}_p \left(\varphi, \frac{E_{T,s}}{\sigma_{y,s}}, A, \zeta \right). \tag{13}$$

Again, we can incorporate the Gibson–Ashby scaling law for the yield stress $\sigma_y/\sigma_{y,s} = C_{\sigma_y} \varphi^{3/2}$ [35], which yields for the first output

$$\frac{\sigma_y}{\sigma_{y,s} \varphi^{3/2}} = \hat{g}_{\sigma_y}^* \left(\frac{E_{T,s}}{\sigma_{y,s}}, A, \zeta \right). \tag{14}$$

Concerning the second output of Equation (13), it is unknown which scaling is appropriate, because the work hardening rate is a slope in the stress–plastic strain diagram. Intuitively, one would follow Equation (14) in favor of an exponent of 3/2. We can answer this question together with PCA and keep the exponent β in the scaling flexible, such that

$$\frac{E_T}{E_{T,s} \varphi^\beta} = \hat{g}_{E_T}^* \left(\frac{E_{T,s}}{\sigma_{y,s}}, A, \zeta \right). \tag{15}$$

Alternatively to Equation (13), only dependent variables are used for normalization of the output

$$\left(\frac{E_T}{\sigma_y}, \nu_p \right) = \tilde{g}_p \left(\varphi, \frac{E_{T,s}}{\sigma_{y,s}}, A, \zeta \right). \tag{16}$$

The choice between the two methods of normalization depends on the potential application. Equation (16) has the advantage that all quantities on the left side are experimentally accessible, such that it could be possible to invert $\tilde{g}_p$ and to obtain some insight into material or structural properties of the nanoporous metal based on macroscopic compression testing.

3.2. Principal Component Analysis

At first glance, principal component analysis (PCA) [47,48] appears to be meaningless for our case, because there are no linear dependencies among the inputs that could be easily eliminated. PCA of linearly independent inputs simply translates the original inputs into a smaller number of components by linear combination. In case that each original input carries important information, this leads to a loss of information and to an increase in the predicted error in a subsequent MLP regression. In contrast, a successful reduction to a fewer number of components without a substantial increase in the prediction uncertainty shows that there is a potential for the reduction of the dimensionality of the problem and, furthermore, it delivers a feeling for the number of inputs that can be removed.

The advantage of PCA is that the data can be quickly analyzed, and it becomes clear which elements of a relationship are the promising candidates for a deeper analysis.

Because the equations for elasticity can be easily visualized, Equations (11) and (12) are omitted here. The mapping of PCA with an MLP regression of Equation (14) is shown in Figure 5a. For these regressions, consistently 10 neurons in a single hidden layer were used. The results for 3 components corresponded to the dimensionality of the raw input data and reproduced the accuracy of the MLP prediction without PCA, validating that no information was lost by the transformation. With reduction of the components, computed mean values of the absolute prediction error were 0.121, 0.221, and 0.342 for 3, 2, and 1 components, respectively. As can be seen from the inserted plot (orange), the error doubled with each component that was reduced. The scatter plot in Figure 5a suggests to visualize the data in form of a 3D plot, where a parametrization with one of the three inputs is required, which is presented in Section 3.4.

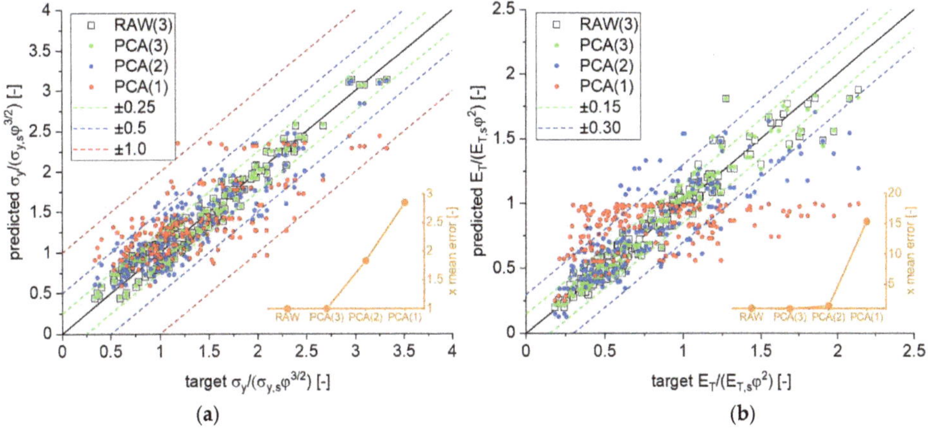

Figure 5. Result of PCA followed by MLP regression with a single hidden layer of 10 neurons for a decreasing number of components: (**a**) scaled yield stress, Equation (14); (**b**) scaled work hardening rate, Equation (15).

A first investigation of Equation (15) with 3 components and an exponent $\beta = 3/2$ similar to Equation (14) led to two main groups in the scatter plot (not shown), which could be combined in a narrow scatter band by changing the exponent to $\beta = 2$ (black open boxes in Figure 5b. Thus, the data suggested that the work hardening rate should be scaled in the same way as the Young's modulus. Using this exponent, we obtained mean values of the absolute error of 0.070, 0.177, and 0.317 for 3, 2, and 1 components, respectively.

Next, it was of interest to quantify the highest possible reduction of arguments of $\widetilde{g}_p$ in Equation (16). The more significant the outcome is, the better are the chances for deriving a relationship that can potentially also be solved with respect to one of the arguments. This would be a valuable aid in accessing local structural or mechanical properties from comparably simple macroscopic tests. Figure 6 presents the outcome of a PCA of $\widetilde{g}_p$ followed by MLP. The PCA was first applied simultaneously to both ligament shapes G_{21} and G_{33}. Each dimensionless parameter on the left side of Equation (16) is individually evaluated.

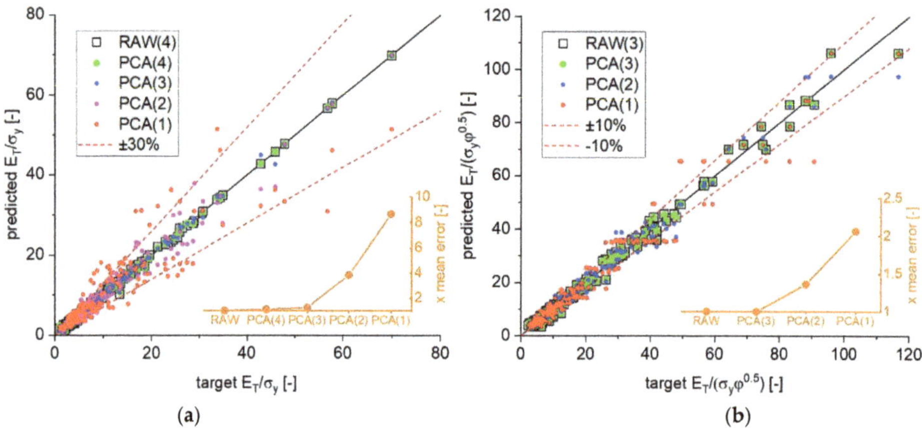

Figure 6. Quality of MLP regression after PCA with decreasing number of components. (**a**) E_T/σ_y, Equation (16) with ≤ 4 components and (**b**) $E_T/\left(\sigma_y\varphi^{1/2}\right)$ scaled according to Equation (18) with ≤ 3 components.

The importance of the correct scaling was demonstrated for the analysis of Equation (16). The results for an output in the form of the ratio E_T/σ_y, shown in Figure 6a, revealed that the argument of Equation (16) could not be easily reduced without adding considerable error. The reduction from 3 to 2 components led to a separation into two branches (reddish colors) that resulted from the two solid fractions. Further reduction did not change the result much.

However, dividing Equation (15) by Equation (14) for $\beta = 2$ yields

$$\frac{E_T\varphi^{3/2}}{\sigma_y\varphi^2}\frac{\sigma_{y,s}}{E_{T,s}} = \widetilde{g}_p^*\left(\frac{E_{T,s}}{\sigma_{y,s}}, A, \zeta\right), \tag{17}$$

which can be rewritten as

$$\frac{E_T}{\sigma_y\varphi^{1/2}} = g_p^*\left(\frac{E_{T,s}}{\sigma_{y,s}}, A, \zeta\right). \tag{18}$$

For this type of scaling, shown in Figure 6b, PCA delivered almost a perfect match, independent of the number of components, suggesting that the argument of Equation (18) can be reduced to a single component. Because the output of Equation (18) can be expected to mainly depend on the corresponding ratio $E_{T,s}/\sigma_{y,s}$, the visualization can right away move to a 2D scatter plot of $E_T/(\sigma_y\varphi^{1/2})$ versus this quantity.

In contrast to the output E_T/σ_y, the plastic Poisson's ratio ν_p showed a large scatter that is almost invariant to the number of components (not shown). Therefore, no further reduction of the dimensionality is possible for this parameter. This is further discussed along with the visualization of the data in Section 3.4.

3.3. Macroscopic Elastic Properties

The dependencies for the elastic properties according to Equations (11) and (12) are visualized in Figures 7a and 8. In these Figures, the randomly distributed simulation data are shown as spheres, and the predictions of the MLP regressions are shown as 3D contour plots. As can be seen from Figure 7, the scaling of Young's modulus removes most of the effect stemming from the solid fraction, such that g_E^* can be written as dependence of only two structural parameters A and ζ. The surfaces approximating the individual solid fractions are slightly shifted in the lower regions and intersect at $E/(E_s\varphi^2) \sim 1.5$.

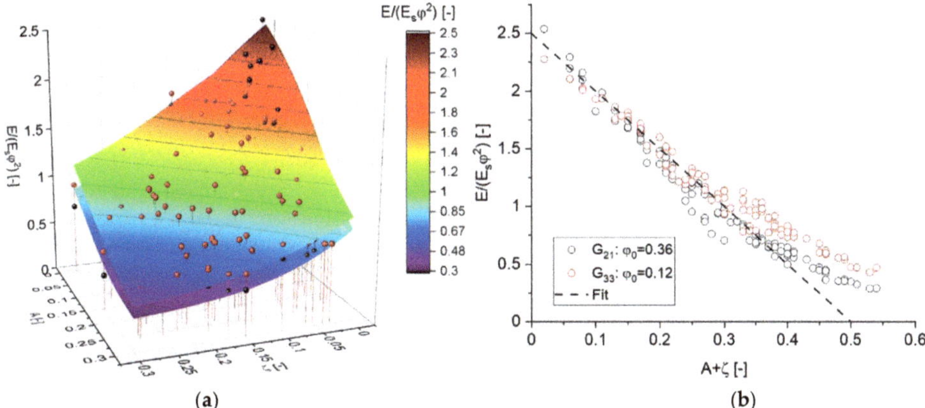

(a)   (b)

Figure 7. Approximation of the simulation results for the scaled Young's modulus $E/(E_s\varphi^2)$. Black and red spheres represent data from ligament shapes G_{21} and G_{33}, respectively, and MLP regressions are shown as contour plot as function of randomization A and cut fraction ζ: (**a**) 3D plot of Equation (11), (**b**) quality of approximation by Equation (19).

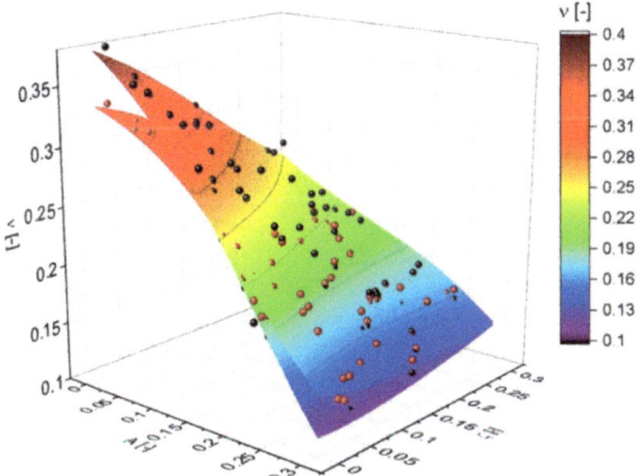

Figure 8. Visualization of the simulation results for the macroscopic Poisson's ratio. Black and red spheres represent data for ligament shapes G_{21} and G_{33}, respectively. MLP regressions of both data sets are shown as contour plots.

Overall, the contour plot in Figure 7a confirms the existing understanding about the effect of A and ζ, which both lower the macroscopic Young's modulus of the ligament network [28]. Additionally, with ζ approaching the percolation threshold, one observes a smooth transition into a horizontal tangent with the x-axis. Interestingly, after considering both parameters in the computation of the solid fraction, the remaining effect is almost identical, as can be seen from the horizontal isolines in Figure 7a. Combining them in the x-axis in Figure 7b reveals where the two data sets start to separate. The dashed line indicates that up to a value of $A + \zeta = 0.3$ the data can be fitted by

$$\frac{E}{E_s\varphi^2} \approx 2.5 - 5(A + \zeta). \tag{19}$$

The macroscopic Poisson's ratio shown in Figure 8 behaves differently. Again, the effect of the randomization A is at least as strong as the effect of the cut fraction ζ. However, in agreement to the findings in [28], the cut fraction of ζ has no effect for $A \sim 0.25$, while it is large for lower values of A and of opposite sign for $A = 0.3$. The Poisson's ratio is only slightly sensitive to φ in the regime of low values of ζ, i.e., for fully connected networks. In summary, as suggested by the PCA, the visualization in Figure 8 confirms that the dimensionality of this relationship cannot be further reduced.

3.4. Macroscopic Plastic Properties

With the outcome of the PCA in mind, Equation (14) is visualized in Figure 9a. Each pair of contour plots correspond to the two solid fractions, the effect of which is captured by the scaling with $\varphi^{3/2}$. In addition to uncertainties and numerical errors, the remaining gap within each pair could be a result, e.g., of torsion that scales with φ and can have some 10% contribution to the deformation as soon as the ligaments are randomized [25]. The effect of $\log(E_{T,s}/\sigma_{y,s})$ is remarkable in all regions of the plot and is around a factor of 2, but also the dependencies of A and ζ are significant. This explains why all three parameters need to be kept for a good representation of Equation (14), as indicated by the PCA. The proper scaling of the work hardening rate with an exponent of $\beta = 2$ is confirmed with Figure 9b, which is in appearance and range of values very close to that of the Young's modulus shown in Figure 7a.

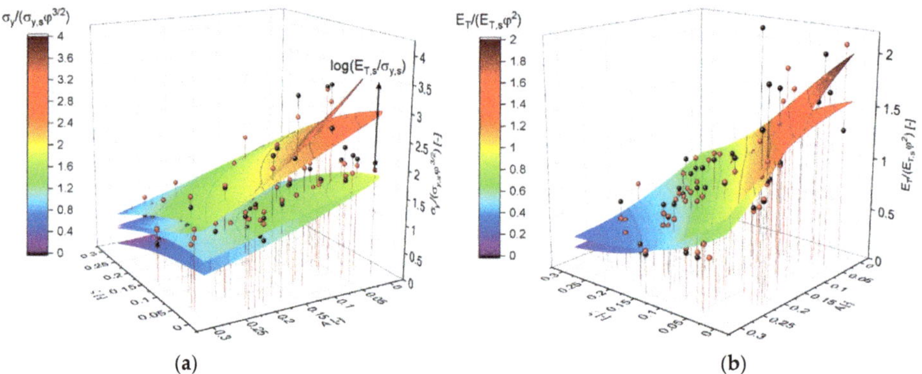

Figure 9. Visualization using MLP regression (shown as contour plot) along with black and red spheres corresponding to the data from ligament shapes G_{21} and G_{33}, respectively. (**a**) Scaled yield stress after Equation (14). Each pair of contour plots represents the two solid fractions. The parametrization of $\log(E_{T,s}/\sigma_{y,s})$ corresponds to the values 0.5 and 2.5. (**b**) Scaled work hardening rate after Equation (15) with an exponent $\beta = 2$, where the inputs for the MLP regression are reduced to A and ζ. The pair of planes represent the two solid fractions.

For obtaining a first impression on the dependence of E_T/σ_y according to Equation (16), the data is visualized in Figure 10. The dependence of $\log(E_T/\sigma_y)$ is clearly the strongest and nicely correlated with $\log(E_{T,s}/\sigma_{y,s})$, whereas A and ζ have no or only a small effect, respectively. Therefore, the data can be plotted as $\log(E_T/\sigma_y)$ versus $\log(E_{T,s}/\sigma_{y,s})$ in a 2D scatter plot. The correlation shown in Figure 11 is linear over the whole range from $0.5 \leq \log(E_{T,s}/\sigma_{y,s}) \leq 2.5$, i.e., from almost perfectly plastic to strongly work hardening materials. The slope is positive, i.e., an increase in the ratio $E_{T,s}/\sigma_{y,s}$ increases the corresponding ratio E_T/σ_y in the macroscopic behavior, which is expected. The scatter around the linear fit is, with few exceptions, ± 0.1, which corresponds to 25% in a linear scaling. This scatter results in part from the additional dependence of ζ, which is visible in a tilt of the contour plots in Figure 10.

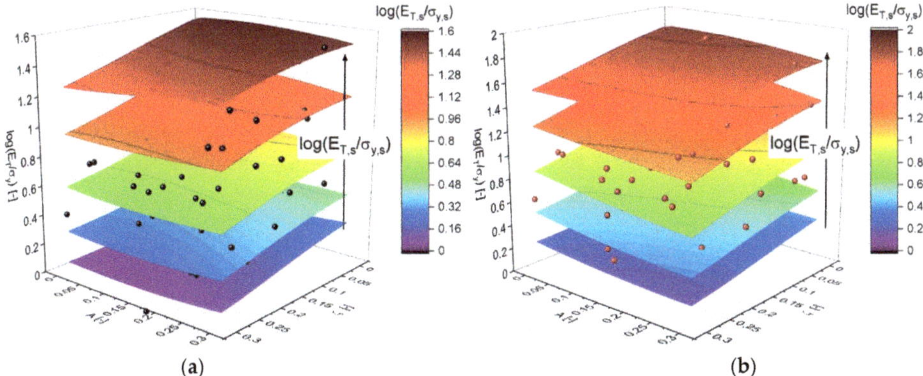

Figure 10. Approximation of the simulation results for the macroscopic plastic properties (spheres) by MLP regression (shown as contour plot) as functions of randomization A and cut fraction ζ: (**a**) G_{21}: $\varphi = 0.12$, (**b**) G_{33}: $\varphi = 0.35$.

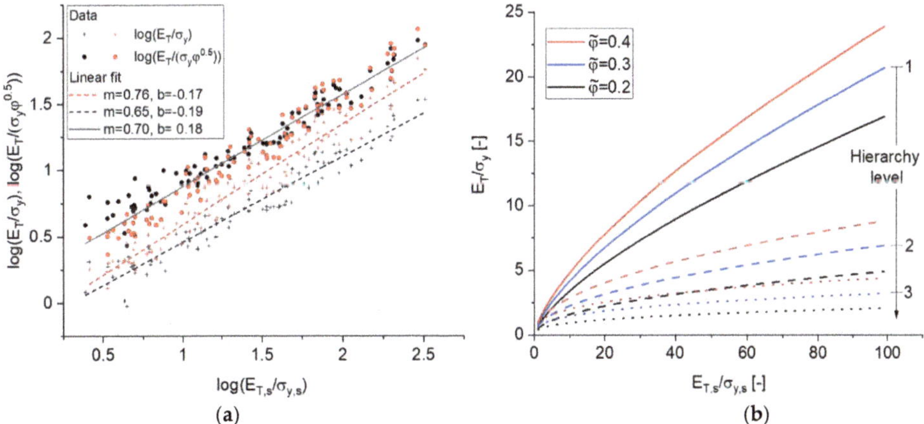

Figure 11. (**a**) Scaled work hardening rate to yield stress reduced to 2D plots for E_T/σ_y—Equation (16) and $E_T/(\sigma_y\varphi^{1/2})$ —Equation (18) for the two data sets G_{21} (black), $\varphi_0 = 0.12$ and G_{33} (red), $\varphi_0 = 0.35$. Linear fits given as $y = mx + b$; (**b**) plot of E_T/σ_y for one and two levels of hierarchy according to Equations (21)–(23), respectively, showing the decreasing influence of $E_{T,s}/\sigma_{y,s}$ with increasing hierarchy level.

The parameters of the linear fits for $\log(E_T/\sigma_y)$ depend on the solid fraction, which could be incorporated, e.g., by linear interpolation between them, because this effect is small compared to the range of $\log(E_T/\sigma_y)$. Alternatively, we can make use of the scaling according to Equation (18), which removes the effect of the solid fraction, such that the two data sets are merged in the scatter plot in Figure 11a for $\log\left(E_T/(\sigma_y\varphi^{1/2})\right)$. The correlation is again linear in the log–log plot and can be fitted with

$$\log(E_T/(\sigma_y\varphi^{\frac{1}{2}})) = 0.18 + 0.7\log(E_{T,s}/\sigma_{y,s}). \tag{20}$$

Equation (20) can be rewritten as

$$\frac{E_T}{\sigma_y} = b\sqrt{\varphi}\left(\frac{E_{T,s}}{\sigma_{y,s}}\right)^{\gamma}, \tag{21}$$

with $b = 1.514$ and $\gamma = 0.7$.

Shi et al. [6] developed scaling laws for the macroscopic Young's modulus and yield stress (in general denoted as property P) for a hierarchically nested network of n levels of the form $P_{net} = b^n P_s \widetilde{\varphi}^{n\beta}$. This results from the recursive application of the Gibson–Ashby scaling law $P_{eff} = b P_s \varphi^\beta$ under the assumption of a strong self-similarity $\varphi_{net} = \widetilde{\varphi}^n$. Here, P_s is the mechanical property of the solid phase, P_{eff} is the effective (homogenized) value, and P_{net} is the result of the net value of P. For the work hardening to yield stress ratio as given by Equation (21), the property itself scales with an exponent P^γ, such that the effective properties on the next hierarchy level are $P_{eff,j} = b\widetilde{\varphi}^\beta P^\gamma_{eff,j-1}$ with $\beta = 0.5$. Therefore, E_T / σ_y for a material with two and three levels of hierarchy is given by

$$\frac{E_T}{\sigma_y} = b^{1+\gamma} \sqrt{\widetilde{\varphi}^{1+\gamma}} \left(\frac{E_{T,s}}{\sigma_{y,s}} \right)^{\gamma^2} \tag{22}$$

and

$$\frac{E_T}{\sigma_y} = b^{1+(1+\gamma)\gamma} \sqrt{\widetilde{\varphi}^{1+(1+\gamma)\gamma}} \left(\frac{E_{T,s}}{\sigma_{y,s}} \right)^{\gamma^3}, \tag{23}$$

respectively. For two levels, the total solid fraction is $\varphi = \widetilde{\varphi}^2$, which ranges from 0.119 to 0.165 [6]. In this case, $\widetilde{\varphi}$ ranges from 0.345 to 0.406, which changes the leading term in Equation (22) by 14%. Because of the exponent $\gamma^2 = 0.49$, a similar effect would require a variation in the material properties $E_{T,s}/\sigma_{y,s}$ by a factor of 1.33. This trend is shown in Figure 11b, for a variation of $E_{T,s}/\sigma_{y,s}$ over two orders of magnitude. If we add a third level of hierarchy, the effect of $E_{T,s}/\sigma_{y,s}$ becomes even smaller $(\gamma^3 = 0.343)$. We can therefore speculate that $E_T/\sigma_y \to 1$ with increasing number of hierarchy levels and E_T/σ_y reduces to a function of $\widetilde{\varphi}$. Section 4 shows that E_T/σ_y is important in the interpretation of the measured hardness.

Finally, the dependency of the plastic Poisson's ratio ν_p in Equation (13) on A, ζ, and $\log(E_{T,s}/\sigma_{y,s})$ is visualized in Figure 12. The MLP regressions shown in Figure 12a reveal that the dependence of ν_p on $\log(E_{T,s}/\sigma_{y,s})$ is significant for low solid fractions, while the effect of the cut fraction ζ is rather small. This changes for high solid fractions shown in Figure 12b, where the effect of $\log(E_{T,s}/\sigma_{y,s})$ is small but the effect of the cut fraction ζ has the same importance as the randomization A. In combination, both parameters can be used to tune the plastic Poisson's ratio over a large range from ~ 0.3 to ~ 0.1. The complex dependency indicates that the multiaxial plastic deformation behavior of nanoporous metals can strongly vary and needs to be determined individually for each microstructure. Additionally, the ligament diameter and surface energy have an important effect on the plastic Poisson's ratio, as shown in [32,49]. These experiments show a comparably large range of values from 0 to 0.2 for increasing ligament size. This range is included in the simulation data shown in Figure 12.

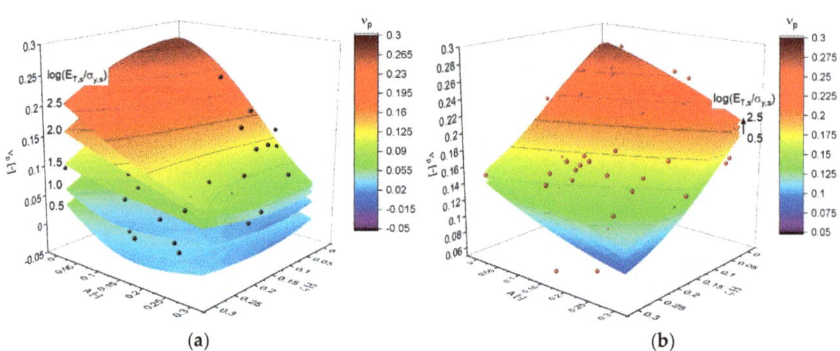

(a)   (b)

Figure 12. Visualization of the plastic Poisson's ratio ν_p in Equation (13) for (a) G_{21}: $\varphi = 0.12$ and (b) G_{33}: $\varphi = 0.35$.

4. Nanoindentation

In this section, the dependence of the hardness was analyzed with respect to the influence of the underlying structural and mechanical properties of the nanofoam. To this end, we performed a dimensionality reduction along the same line as in Section 3 with (i) dimensional analysis, (ii) principal component analysis, and (iii) visualization and modeling of the relationship with a minimum number of inputs.

4.1. Dimensional Analysis

The major output of a nanoindentation experiment was the hardness H, which can be written as

$$H = \overline{H}\left(r_{mid},\ r_{end},\ l,\ E_s,\ v_s,\ \sigma_{y,s},\ E_{T,s},\ A, \zeta\right). \tag{24}$$

As in Section 3.1, we represented the ligament shape defined by r_{mid}, r_{end}, and l by the solid fraction φ and, furthermore, assumed that the hardness is governed by plastic and structural parameters, while the effect of the elastic material parameters of the comparably soft solid phase can be neglected, i.e., $\sigma_{y,s} \ll E_s$ and $E_{T,s} \ll E_s$. This reduces Equation (24) to

$$H = \hat{H}\left(\varphi,\ \sigma_{y,s},\ E_{T,s},\ A, \zeta\right). \tag{25}$$

The hardness mainly scales with the macroscopic yield stress, as this is the case for bulk materials [50]. Hence, writing Equation (25) in dimensionless form and considering that $E_{T,s}/\sigma_{y,s}$ can be replaced by a dependence of E_T/σ_y and φ using Equation (21), this leads to a relationship that includes only macroscopic properties and structural parameters:

$$\frac{H}{\sigma_y} = \hat{H}^*\left(\varphi,\ \frac{E_T}{\sigma_y},\ A, \zeta\right). \tag{26}$$

4.2. Principal Component Analysis

For further reduction of Equation (26), we used the hardness results from the indentation simulations described in Section 2.2 that were carried out with the same parameter sets as the simulations for uniaxial compression in Section 2.1. A PCA of Equation (26), shown in Figure 13, suggested that the four arguments could be reduced to one, when an uncertainty in the predicted H/σ_y from ± 0.1 to ± 0.3 is acceptable. A reduction of the uncertainty to ± 0.2 would already require at least three components. This potential for simplification is also reflected in the factor by which the absolute mean error is increased due to the reduction of the number of components, shown in the insert (orange) in Figure 13. By a reduction to a single component, this error measure is only increased by a factor of 1.3, which is a very low value compared to the results in Section 3.2.

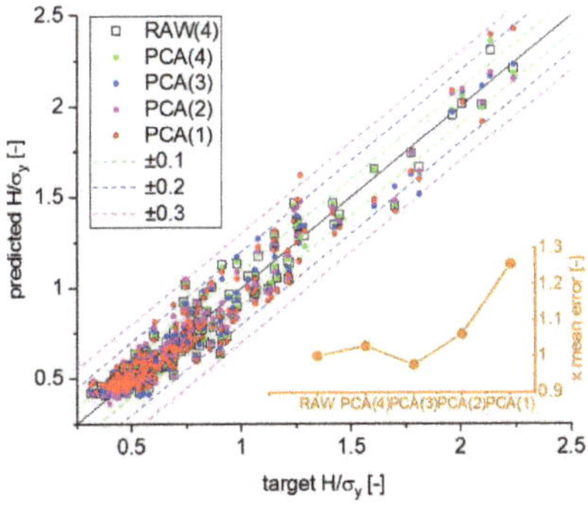

Figure 13. Principal component analysis of the dependence of H/σ_y with respect to structural and macroscopic mechanical properties following Equation (26), combining data sets G_{21} and G_{33}.

4.3. Hardness

The quantitative dependence of the normalized hardness H/σ_y as function of the structural parameters (A, ζ) and macroscopic material properties $\log(E_T/\sigma_y)$ is shown in Figure 14. The MLP regressions are shown as contour plots in Figure 14a,b, confirming that $\log(E_T/\sigma_y)$ is the most important parameter, followed by the randomization A, which has a moderate effect, whereas the effect of the cut fraction ζ can be neglected. In Figure 14c,d, the axis of the cut fraction ζ is replaced by $\log(E_T/\sigma_y)$. A small effect of the randomization A with a negative slope in the low solid fraction data can be expressed by $H/\sigma_y \approx H/\sigma_y\big|_{A=0} - 0.16A$. For $A = 0.3$, this effect is $\Delta H/\sigma_y \leq 0.05$, which corresponds to the uncertainty when structural effects are not taken into account. For solid fractions of typical samples with $\varphi \approx 0.32$, the effect caused by structural disorder is negligible.

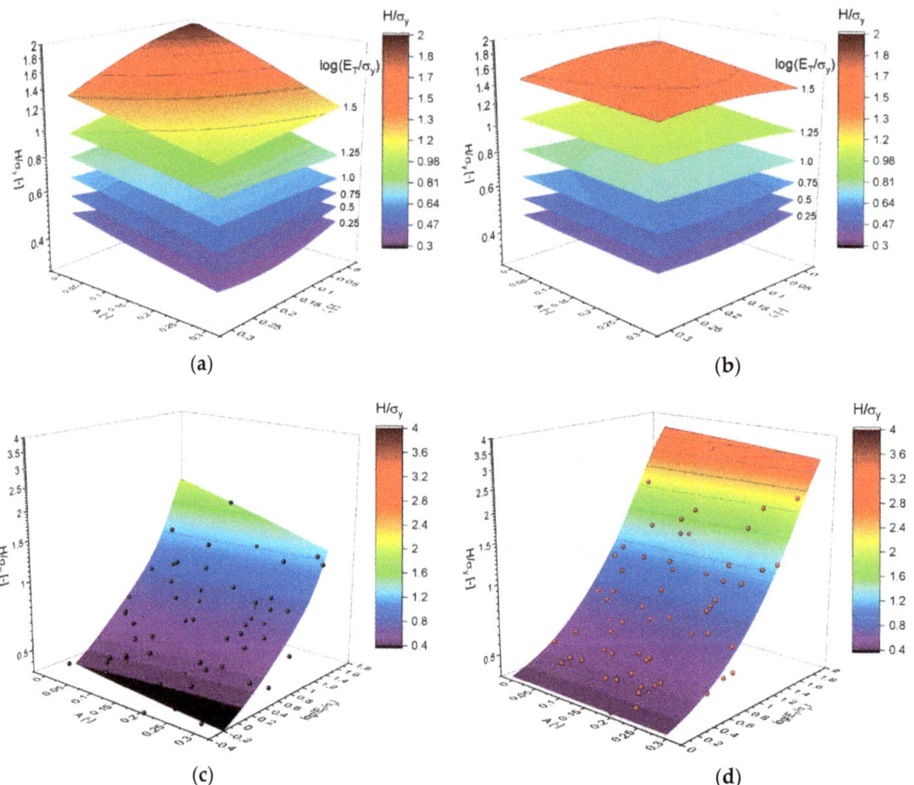

Figure 14. (**a**,**b**) Dependence of the normalized hardness H/σ_y as function of structural properties (A, ζ) and macroscopic material properties $\log(E_T/\sigma_y)$; (**c**,**d**) reduction of dimensionality by elimination of the cut fraction ζ, confirming that the simulation data can be represented by a simple dependence $H/\sigma_y(E_T/\sigma_y)$. Ligament geometries are (**a**,**c**) G_{21}, $\varphi = 0.12$; (**b**,**d**) G_{33}, $\varphi = 0.35$.

We could further reduce the relationship to a 2D scatter plot, shown in Figure 15, which is fitted with a linear relation

$$\frac{H}{\sigma_y} = H_0^* + m_H \frac{E_T}{\sigma_y},\tag{27}$$

where for our data, we obtained $H_0^* = 0.41$ and $m_H = 0.035$. In this Figure, the error bars correspond to the standard deviation of ± 0.11. Both ligament geometries are combined in this plot, which indicates that the dependence Equation (27) is applicable for a broad range of structures and is insensitive to microstructural parameters.

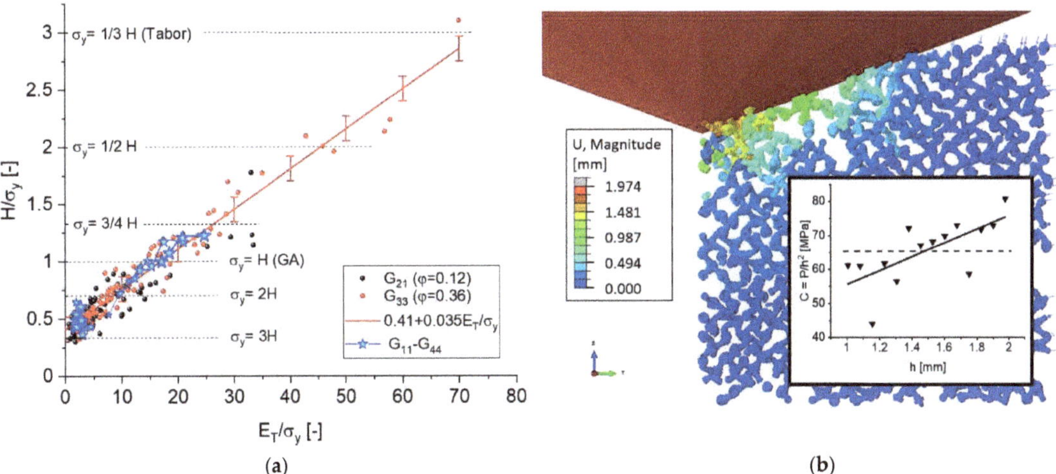

Figure 15. (a) Correlation of H/σ_y with E_T/σ_y for ligament shapes G_{21} ($\varphi_0 = 0.12$) and G_{33} ($\varphi_0 = 0.36$). (b) Deformation shown for a slice of one unit cell of the indented RVE with $A = 0.18$ and $\zeta = 0.26$, for which the stress–strain curve is shown in Figure 2b.

For confirmation, additional simulations for all 16 ligament geometries G_{11} to G_{44} [11,27] were added. These data points, entered as star symbols, represent the combinations of $r_{end}/l \in \{0.231, 0.289, 0.346, 0.404\}$ and $r_{mid}/r_{end} \in \{0.5, 0.75, 1.0, 1.25\}$ for two values $E_{T,s}/\sigma_{y,s} \in \{3.16, 50.0\}$. This adds 32 simulations that provide an insight into possible dependencies of the ligament shape and solid fraction. The results are added in Figure 15 as blue stars, where the blue curves connect simulation results of constant r_{end}/l and the line thickness increases with the value of r_{end}/l. All results are within the scatter of the random simulations for geometries G_{21} and G_{33}, confirming that Equation (27) holds for all ligament shapes and solid fractions within the given scatter band. While for $E_{T,s}/\sigma_{y,s} = 3.16$ the data scatter around a spot in the lower left area of the plot, the results for $E_{T,s}/\sigma_{y,s} = 50$ show that with increasing r_{mid}/r_{end} and solid fraction φ the data points systematically move towards larger ratios H/σ_y. The same applies to the random data, when we compare the range of values for the geometries $G_{21}(\varphi_0 = 0.12)$ and G_{33} ($\varphi_0 = 0.35$) in black and red, respectively.

Despite the common assumption for foams $H/\sigma_y = 1$, it seems reasonable that a porous material tends towards a bulk solid for a high solid fraction. However, it is difficult to understand that the hardness can fall below the macroscopic yield stress. This could be caused by the way the macroscopic stress–strain behavior has been translated into the material parameters E_T and σ_y, as shown in Figure 2b. The macroscopic yield stress is read from the linear fit of the stress–strain curve for plastic strains $> 1\%$. This procedure removes initial nonlinearities that could be interpreted as microplasticity. However, microplasticity does not exist in our continuum model; therefore, the true yield stress is usually lower than that determined from the linear fit. In the example shown in Figure 2b, the measured yield stress determined from the linear hardening model was 7.3 MPa, whereas at 0.2% plastic strain, the stress reached a value of only 5 MPa. For $H = 3.2$ MPa, the ratio H/σ_y then changed from 0.43 to 0.64, if the yield stress at 0.2% plastic strain was used. This explains in part why the hardness can be lower than the yield stress.

It could be speculated that another contribution might stem from the reduction of the connectivity of the ligament network, as shown in Figure 15b. The RVE is characterized by a cut fraction of $\zeta = 0.26$, i.e., almost a third of the ligaments in the RVE are broken. This leads to large pores, which become comparable to the indentation depth and the contact radius. When a microstructural length and the indentation depth are of the same order, the simulation shows a size effect. For 3D networks, this problem becomes relevant when approaching the percolation threshold and is difficult to solve [51]. An increase in the normalized indentation load P/h^2, shown as insert in Figure 15b, apparently confirms this effect. However, one would then also expect a systematic bias in the data in the form of a dependence of the cut fraction, i.e., $H/\sigma_y \to 1$ for $\zeta \to 0$, but such a trend is not present in Figure 14a,b. It is therefore possible that values $H/\sigma_y < 1$ exist. Because this has important implications on the interpretation of hardness data of foams in general, an in-depth investigation should be the scope of future work.

5. Summary and Conclusions

Nanoporous metals with their complex microstructure represent an ideal candidate for method developments that combine data and AI. With a few parameters controlling the sample preparation, it is possible to tune the microstructure and macroscopic mechanical properties within a large design space. This includes, among others, the solid fraction, ligament size, and the connectivity density. It has been recently demonstrated that the versatile dealloying process allows hierarchically organized nanoporous metals with superior macroscopic properties to be produced compared to those with only one hierarchy level [6]. Via the microstructure, it is possible to tune the macroscopic properties, such as Young's modulus, yield strength, elastic and plastic Poisson's ratio, and hardness in wide ranges. This makes this class of materials not only attractive for various applications, such as sensing or actuation in combination with light weighting, but it is also an ideal science case for the demonstration of the capabilities of dimensionality reduction methods.

To this end, the generation of ~200 data sets for macroscopic compression and nanoindentation was realized with the help of an efficient FE-beam modeling technique. The parameter space consists of five independent inputs (microstructure, material parameters) and six dependent outputs (macroscopic compression behavior and hardness). It was systematically analyzed in three steps by means of a dimensional analysis including a priori knowledge about the problem at hand, principal component analysis, and visualization. In the latter two steps, machine learning served as key for analyzing the existence and quality of approximations on the presented data sets.

From the outcome, we conclude that, independent of the size of the data set, it is always recommendable to start with a dimensional analysis. This ensures that the analyzed dependency is formulated in a physically reasonable manner and it allows the dimensionality of the problem to be reduced by usually two quantities in quasi static mechanics or by three for dynamic problems. At this stage, it is advisable to incorporate *a priori* knowledge from the literature or by reasoning, which can further simplify the problem considerably. How well this has been done and by how many components the dependency can be further reduced can be easily tested by machine learning in combination with principal component analysis. If no deeper understanding is needed, the outcome in the form of a black box would already be a sufficient computer model of the relationship hidden in the presented data.

Deeper insight can be gained by visualization, which is also supported by machine learning. Here, the multilayer perceptron first approximates the design space from the randomly distributed data and then is applied for continuous mapping along selected inputs in the form of contour plots. This serves the validation of the previous steps as well as for a better understanding of the quantitative dependence of a specific output, e.g., the hardness to yield stress ratio, of inputs, e.g., the randomization or mechanical properties. In this way, the major dependences can be identified from a limited number of data and unimportant inputs can be eliminated. Furthermore, one obtains a measure for the uncertainty due to ignored inputs that have a non-negligible effect.

For the scientific case at hand, which is the microstructure–property relationship of nanoporous metals, there are several important findings, which are applicable not only to Au but to any metal, as long as it can be described with the chosen elastic–plastic material behavior and microstructure. Our analysis showed that the ratio of the work hardening rate to the yield stress $E_{T,s}/\sigma_{y,s}$ represents a key property that can mapped to the corresponding macroscopic ratio E_T/σ_y in a log–log scaling. It is therefore possible to invert this relationship for measured macroscopic behavior, which allows one to gain an important insight in the amount of work hardening present in the solid phase, relative to its yield stress. Work hardening implies storage of defects in the nanoscaled ligaments, and its existence has been a matter of debate. The derived relationship can help to quantitatively underpin speculations that are in favor [52] or contradict [13,23] the mechanistic model of dislocation starvation in nanosized metallic objects simply by translating the macroscopic test data into those of the solid phase.

In addition to the known Gibson–Ashby scaling laws for Young's modulus and yield strength, one for the work hardening rate is added, which uses the same exponent of 2 as the Young's modulus. This is unexpected, because the work hardening rate is a slope defined by two flow stresses at different plastic strains and the yield stress, which is one of them, scales with an exponent of 1.5. Additionally, the appearance and range of the relationship as functions of randomization and cut fraction are very similar to that of the Young's modulus.

Another important finding is the linear relation between H/σ_y and E_T/σ_y. The common assumption that for hardness testing of foams $H = \sigma_y$ [35], which is also used in the interpretation of nanoindentation of np-Au [37,38], turned out to be a special case for $E_T/\sigma_y \sim 17$. The range of the H/σ_y data is surprisingly large and exceeds the common values for porous and bulk solids of 1 and 3, respectively, towards lower values: A large number of data are within the range $0.5 \leq H/\sigma_y \leq 1$. This can in part be explained by how the macroscopic stress–plastic strain curve is modeled. Another reason could be a size effect that results from large pores for samples with very low connectivity, but it appears that still values of $H/\sigma_y < 1$ exist. Because this has important implications on the interpretation of hardness data, an in-depth investigation will be the scope of future work.

Finally, for hierarchic materials with a nested network [6], our results suggest that the effect of $E_{T,s}/\sigma_{y,s}$ becomes small or even negligible with respect to E_T/σ_y. With increasing levels of hierarchy, it can be expected that the normalized hardness H/σ_y changes from a dependence of E_T/σ_y, which holds for a common nanoporous metal, towards a dependence of the solid fraction φ.

Funding: Support was provided by Deutsche Forschungsgemeinschaft—Project Number 192346071—SFB 986 "Tailor-Made Multi-Scale Materials Systems: M3," project B4.

Institutional Review Board Statement: Not applicable.

Informed Consent Statement: Not applicable.

Data Availability Statement: The data sets analyzed in this work are available via the repository TORE (https://tore.tuhh.de/ accessed on 3 April 2021) published under https://doi.org/10.15480/3 36.3411.

Acknowledgments: N.H. acknowledges Ilona Ryl, who carried out preliminary FE simulations with a hybrid FE-beam and FE-solid modeling technique.

Conflicts of Interest: The authors declare no conflict of interest.

References

1. Lilleodden, E.T.; Voorhees, P.W. On the topological, morphological, and microstructural characterization of nanoporous metals. *MRS Bull.* **2018**, *43*, 20–26. [CrossRef]
2. Weissmüller, J.; Sieradzki, K. Dealloyed nanoporous materials with interface-controlled behavior. *MRS Bull.* **2018**, *43*, 14–19. [CrossRef]

3. Jin, H.-J.; Weissmüller, J.; Farkas, D. Mechanical response of nanoporous metals: A story of size, surface stress, and severed struts. *MRS Bull.* **2018**, *43*, 35–42. [CrossRef]

4. Biener, J.; Hodge, A.M.; Hayes, J.R.; Volkert, C.A.; Zepeda-Ruiz, L.A.; Hamza, A.V.; Abraham, F.F. Size effects on the mechanical behavior of nanoporous Au. *Nano Lett.* **2006**, *6*, 2379–2382. [CrossRef] [PubMed]

5. Mameka, N.; Wang, K.; Markmann, J.; Lilleodden, E.T.; Weissmüller, J. Nanoporous Gold—Testing Macro-scale Samples to Probe Small-scale Mechanical Behavior. *Mater. Res. Lett.* **2016**, *4*, 27–36. [CrossRef]

6. Shi, S.; Li, Y.; Ngo-Dinh, B.-N.; Markmann, J.; Weissmüller, J. Scaling behavior of stiffness and strength of hierarchical network nanomaterials. *Science* **2021**, *371*, 1026–1033. [CrossRef]

7. Li, Y.; Dinh Ngô, B.-N.; Markmann, J.; Weissmüller, J. Topology evolution during coarsening of nanoscale metal network structures. *Phys. Rev. Mater.* **2019**, *3*. [CrossRef]

8. Ngô, B.-N.D.; Roschning, B.; Albe, K.; Weissmüller, J.; Markmann, J. On the origin of the anomalous compliance of dealloying-derived nanoporous gold. *Scr. Mater.* **2017**, *130*, 74–77. [CrossRef]

9. Richert, C.; Huber, N. A Review of Experimentally Informed Micromechanical Modeling of Nanoporous Metals: From Structural Descriptors to Predictive Structure-Property Relationships. *Materials* **2020**, *13*, 3307. [CrossRef]

10. Richert, C.; Huber, N. Skeletonization, Geometrical Analysis, and Finite Element Modeling of Nanoporous Gold Based on 3D Tomography Data. *Metals* **2018**, *8*, 282. [CrossRef]

11. Richert, C.; Odermatt, A.; Huber, N. Computation of Thickness and Mechanical Properties of Interconnected Structures: Accuracy, Deviations, and Approaches for Correction. *Front. Mater.* **2019**, *6*, 352. [CrossRef]

12. Farkas, D.; Caro, A.; Bringa, E.; Crowson, D. Mechanical response of nanoporous gold. *Acta Mater.* **2013**, *61*, 3249–3256. [CrossRef]

13. Ngô, B.-N.D.; Stukowski, A.; Mameka, N.; Markmann, J.; Albe, K.; Weissmüller, J. Anomalous compliance and early yielding of nanoporous gold. *Acta Mater.* **2015**, *93*, 144–155. [CrossRef]

14. Farkas, D.; Stuckner, J.; Umbel, R.; Kuhr, B.; Demkowicz, M.J. Indentation response of nanoporous gold from atomistic simulations. *J. Mater. Res.* **2018**, *33*, 1382–1390. [CrossRef]

15. Saane, S.S.R.; Mangipudi, K.R.; Loos, K.U.; de Hosson, J.T.M.; Onck, P.R. Multiscale modeling of charge-induced deformation of nanoporous gold structures. *J. Mech. Phys. Solids* **2014**, *66*, 1–15. [CrossRef]

16. Soyarslan, C.; Bargmann, S.; Pradas, M.; Weissmüller, J. 3D stochastic bicontinuous microstructures: Generation, topology and elasticity. *Acta Materialia* **2018**, *149*, 326–340. [CrossRef]

17. Roberts, A.P.; Garboczi, E.J. Elastic properties of model random three-dimensional open-cell solids. *J. Mech. Phys. Solids* **2002**, *50*, 33–55. [CrossRef]

18. Gong, L.; Kyriakides, S.; Jang, W.-Y. Compressive response of open-cell foams. Part I: Morphology and elastic properties. *Int. J. Solids Struct.* **2005**, *42*, 1355–1379. [CrossRef]

19. Kanaun, S.; Tkachenko, O. Representative volume element and effective elastic properties of open cell foam materials with random microstructures. *JOMMS* **2007**, *2*, 1607–1628. [CrossRef]

20. Jang, W.-Y.; Kraynik, A.M.; Kyriakides, S. On the microstructure of open-cell foams and its effect on elastic properties. *Int. J. Solids Struct.* **2008**, *45*, 1845–1875. [CrossRef]

21. Luxner, M.H.; Stampfl, J.; Pettermann, H.E. Finite element modeling concepts and linear analyses of 3D regular open cell structures. *J. Mater. Sci.* **2005**, *40*, 5859–5866. [CrossRef]

22. Luxner, M.H.; Stampfl, J.; Pettermann, H.E. Numerical simulations of 3D open cell structures—influence of structural irregularities on elasto-plasticity and deformation localization. *Int. J. Solids Struct.* **2007**, *44*, 2990–3003. [CrossRef]

23. Huber, N.; Viswanath, R.N.; Mameka, N.; Markmann, J.; Weißmüller, J. Scaling laws of nanoporous metals under uniaxial compression. *Acta Mater.* **2014**, *67*, 252–265. [CrossRef]

24. Roschning, B.; Huber, N. Scaling laws of nanoporous gold under uniaxial compression: Effects of structural disorder on the solid fraction, elastic Poisson's ratio, Young's modulus and yield strength. *J. Mech. Phys. Solids* **2016**, *92*, 55–71. [CrossRef]

25. Jiao, J.; Huber, N. Deformation mechanisms in nanoporous metals: Effect of ligament shape and disorder. *Comput. Mater. Sci.* **2017**, *127*, 194–203. [CrossRef]

26. Jiao, J.; Huber, N. Effect of nodal mass on macroscopic mechanical properties of nanoporous metals. *Int. J. Mech. Sci.* **2017**, *134*, 234–243. [CrossRef]

27. Odermatt, A.; Richert, C.; Huber, N.; Odermatt, A.; Richert, C.; Huber, N. Prediction of elastic-plastic deformation of nanoporous metals by FEM beam modeling: A bottom-up approach from ligaments to real microstructures. *Mater. Sci. Eng. A* **2020**, *791*, 139700. [CrossRef]

28. Huber, N. Connections Between Topology and Macroscopic Mechanical Properties of Three-Dimensional Open-Pore Materials. *Front. Mater.* **2018**, *5*, 5801. [CrossRef]

29. Dassault Systemes. *Abaqus 3DEXPERIENCE*; Dassault Systemes SIMULIA Corp.: Johnston, RI, USA, 2020.

30. Mangipudi, K.R.; Epler, E.; Volkert, C.A. Topology-dependent scaling laws for the stiffness and strength of nanoporous gold. *Acta Materialia* **2016**, *119*, 115–122. [CrossRef]

31. Hu, K.; Ziehmer, M.; Wang, K.; Lilleodden, E.T. Nanoporous gold: 3D structural analyses of representative volumes and their implications on scaling relations of mechanical behaviour. *Philos. Mag.* **2016**, *96*, 3322–3335. [CrossRef]

32. Lührs, L.; Zandersons, B.; Huber, N.; Weissmüller, J. Plastic Poisson's Ratio of Nanoporous Metals: A Macroscopic Signature of Tension-Compression Asymmetry at the Nanoscale. *Nano Lett.* **2017**, *17*, 6258–6266. [CrossRef]

33. Liu, L.-Z.; Jin, H.-J. Scaling equation for the elastic modulus of nanoporous gold with "fixed" network connectivity. *Appl. Phys. Lett.* **2017**, *110*, 211902. [CrossRef]

34. Huber, N.; Nix, W.D.; Gao, H. Identification of elastic-plastic material parameters from pyramidal indentation of thin films. *Proc. R. Soc. Lond. A* **2002**, *458*, 1593–1620. [CrossRef]

35. Gibson, L.J.; Ashby, M.F. *Cellular Solids: Structure & Properties*, 1st ed.; Pergamon Press: Oxford, UK, 1988; ISBN 0-08-036607-4.

36. Fischer-Cripps, A.C. *Nanoindentation*, 2nd ed.; Springer: New York, NY, USA, 2004; ISBN 978-1-4757-5943-3.

37. Biener, J.; Hodge, A.M.; Hamza, A.V.; Hsiung, L.M.; Satcher, J.H. Nanoporous Au: A high yield strength material. *J. Appl. Phys.* **2005**, *97*, 24301. [CrossRef]

38. Hodge, A.M.; Biener, J.; Hayes, J.R.; Bythrow, P.M.; Volkert, C.A.; Hamza, A.V. Scaling equation for yield strength of nanoporous open-cell foams. *Acta Materialia* **2007**, *55*, 1343–1349. [CrossRef]

39. Gibbings, J.C. *Dimensional Analysis*; Springer: London, UK, 2011; ISBN 1849963177.

40. Huber, N.; Tsakmakis, C. A neural network tool for identifying the material parameters of a finite deformation viscoplasticity model with static recovery. *Comput. Methods Appl. Mech. Eng.* **2001**, *191*, 353–384. [CrossRef]

41. Tyulyukovskiy, E.; Huber, N. Identification of viscoplastic material parameters from spherical indentation data: Part I. Neural networks. *J. Mater. Res.* **2006**, *21*, 664–676. [CrossRef]

42. Pedregosa, F.; Varoquaux, G.; Gramfort, A.; Michel, V.; Thirion, B.; Grisel, O.; Blondel, M.; Prettenhofer, P.; Weiss, R.; Dubourg, V.; et al. Scikit-learn: Machine Learning in Python. *J. Mach. Learn. Res.* **2011**, *12*, 2825–2830.

43. Bock, F.E.; Aydin, R.C.; Cyron, C.J.; Huber, N.; Kalidindi, S.R.; Klusemann, B. A Review of the Application of Machine Learning and Data Mining Approaches in Continuum Materials Mechanics. *Front. Mater.* **2019**, *6*, 443. [CrossRef]

44. Huber, N.; Kalidindi, S.R.; Klusemann, B.; Cyron, C.J. (Eds.) *Machine Learning and Data Mining in Materials Science*; Frontiers Media SA: Lausanne, Switzerland, 2020; ISBN 9782889636518.

45. Chinesta, F.; Cueto, E.; Abisset-Chavanne, E.; Duval, J.L.; Khaldi, F.E. Virtual, Digital and Hybrid Twins: A New Paradigm in Data-Based Engineering and Engineered Data. *Arch. Computat. Methods Eng.* **2020**, *27*, 105–134. [CrossRef]

46. Gibson, L.J.; Ashby, M.F. The mechanics of three-dimensional cellular materials. *Proc. R. Soc. Lond. A* **1982**, *382*, 43–59. [CrossRef]

47. Lee, J.A.; Verleysen, M. (Eds.) *Nonlinear Dimensionality Reduction*; Springer Science+Business Media LLC: New York, NY, USA, 2007; ISBN 9780387393513.

48. Yun, M.; Argerich, C.; Cueto, E.; Duval, J.L.; Chinesta, F. Nonlinear Regression Operating on Microstructures Described from Topological Data Analysis for the Real-Time Prediction of Effective Properties. *Materials* **2020**, *13*, 2335. [CrossRef] [PubMed]

49. Lührs, L.; Soyarslan, C.; Markmann, J.; Bargmann, S.; Weissmüller, J. Elastic and plastic Poisson's ratios of nanoporous gold. *Scr. Mater.* **2016**, *110*, 65–69. [CrossRef]

50. Tabor, D. *The Hardness of Metals*; Oxford University Press: Oxford, UK, 2000; ISBN 9780198507765.

51. Nachtrab, S.; Kapfer, S.C.; Arns, C.H.; Madadi, M.; Mecke, K.; Schröder-Turk, G.E. Morphology and linear-elastic moduli of random network solids. *Adv. Mater. Weinheim.* **2011**, *23*, 2633–2637. [CrossRef]

52. Greer, J.R.; Nix, W.D. Nanoscale gold pillars strengthened through dislocation starvation. *J. Mater. Res.* **2006**, *73*, 4125. [CrossRef]

 *materials*

Article

Nonlinear Regression Operating on Microstructures Described from Topological Data Analysis for the Real-Time Prediction of Effective Properties

Minyoung Yun [1], **Clara Argerich** [1], **Elias Cueto** [2], **Jean Louis Duval** [3] **and Francisco Chinesta** [1,3,*]

[1] PIMM Laboratory & ESI Group Chair, Arts et Métiers Institute of Technology, CNRS, Cnam, HESAM Université, 151 boulevard de l'Hôpital, 75013 Paris, France; minyoung.yun@ensam.eu (M.Y.); clara.argerich_martin@ensam.eu (C.A.)

[2] Aragon Institute of Engineering Research, Universidad de Zaragoza, 50009 Zaragoza, Spain; ecueto@unizar.es

[3] ESI Group, Bâtiment Seville, 3bis rue Saarinen, 50468 Rungis, France; Jean-Louis.Duval@esi-group.com

* Correspondence: Francisco.Chinesta@ensam.eu

Received: 25 April 2020; Accepted: 13 May 2020; Published: 19 May 2020

Abstract: Real-time decision making needs evaluating quantities of interest (QoI) in almost real time. When these QoI are related to models based on physics, the use of Model Order Reduction techniques allows speeding-up calculations, enabling fast and accurate evaluations. To accommodate real-time constraints, a valuable route consists of computing parametric solutions—the so-called computational vademecums—that constructed off-line, can be inspected on-line. However, when dealing with shapes and topologies (complex or rich microstructures) their parametric description constitutes a major difficulty. In this paper, we propose using Topological Data Analysis for describing those rich topologies and morphologies in a concise way, and then using the associated topological descriptions for generating accurate supervised classification and nonlinear regression, enabling an almost real-time evaluation of QoI and the associated decision making.

Keywords: machine learning; data-driven mechanics; TDA; *Code2Vect*; nonlinear regression; effective properties; microstructures

---

1. Introduction

Recently, industry is experiencing a new revolution. In the past, product design, as well as their associated manufacturing processes, were based on the use of nominal models, nominal loadings (in their broadest sense), and a small amount of data for calibrating those models, with the product performance as a design target.

Very recently, predictions enabling real-time decision-making targeting zero defects in processing and zero unexpected faults in operation, were needed everywhere within the Internet of Things (IoT) paradigm, on the work-floor (smart processes), in the city (autonomous systems and smart-city), at the nation level (e.g., smart nation), etc., i.e., anywhere where engineering designs operate.

In those circumstances, the use of traditional simulation-based engineering (SBE) that was the major protagonist of 20th century engineering, is not anymore a valuable option due to three main reasons: (i) models become sometimes crude approximations of the observed reality; (ii) assimilating data enabling the continuous calibration of the models in operation remains difficult to perform under the stringent real-time constraint; and (iii) the real-time simulation of those extremely complex mathematical models needs alternative techniques to those commonly employed in traditional SBE.

It was at the beginning of the XXI century that two new revolutions in the domain of digital engineering emerged.

1.1. Model Order Reduction

Advances in applied mathematics, computer science (high-performance computing) and computational mechanics met to give rise to a diversity of Model Order Reduction (MOR) techniques [1]. These techniques do not reduce or modify the model, they simply reduce the complexity of its resolution and thus transform a complex and time-consuming calculation, into a real-time response while maintaining precision. These new techniques have completely altered traditional approaches of simulation, optimization, inverse analysis, control and uncertainty propagation, all them operating under the stringent real-time constraint.

In a few words, when approximating the solution $u(\mathbf{x}, t)$ of a given Partial Differential Equation (PDE), the multipurpose finite element method assumes an approximation

$$u(\mathbf{x}, t) = \sum_{i=1}^{N} U_i(t) N_i(\mathbf{x}), \tag{1}$$

where U_i represents the value of the unknown field at node i and $N_i(\mathbf{x})$ is the associated shape function. When N (the number of nodes) increases the solution process becomes cumbersome.

POD-based model order reduction learns offline the most adequate (in a given sense) reduced approximation basis $\{\phi_1(\mathbf{x}), \cdots, \phi_R(\mathbf{x})\}$, and project the solution in it

$$u(\mathbf{x}, t) \approx \sum_{i=1}^{R} \xi_i(t) \phi_i(\mathbf{x}), \tag{2}$$

where now, the complexity scales with R instead of N, with R $\ll$ N in general.

The so-called Proper Generalized Decomposition (PGD from now on) goes a step forward and assume a general approximation

$$u(\mathbf{x}, t) \approx \sum_{i=1}^{M} T_i(t) X_i(\mathbf{x}), \tag{3}$$

where now both the space and time functions, $X_i(\mathbf{x})$ and $T_i(t)$ respectively, are computed during the solution process.

A particularly appealing extension of the just introduced space-time separated representation consists of the space-time-parameter separated representation leading to the a so-called *computational vademecum* that expresses the solution of a parametrized PDE from [2,3]

$$u(\mathbf{x}, t, \mu_1, \ldots, \mu_Q) \approx \sum_{i=1}^{M} X_i(\mathbf{x}) T_i(t) \prod_{j=1}^{Q} M_i^j(\mu_j), \tag{4}$$

where μ_j, $j = 1, \ldots, Q$, represent the model parameters. Once constructed off-line that parametric solution (4), it offers under very stringent real-time constraints—in the order of milliseconds—simulation, optimization, inverse analysis, uncertainty propagation and simulation-based control, to cite a few. Thus, at the beginning of the third millennium a real-time dialogue with physics no longer seemed to be the domain of the impossible.

PGD-based techniques have been widely considered for the real-time simulation and decision-making in a variety of problems of industrial relevance. However, prior to use it, one must extract the parameters to be included as extra-coordinates in the problem statement, and then included in the parametric representation of its solution. In the case of morphological and topological descriptions, as considered later in the present work, the extraction of the adequate parametrization represents the most difficult task. Some attempts of combing PGD-based MOR and manifold learning [4] were addressed in [5–8].

1.2. Engineered Artificial Intelligence

Data bursts within engineering disciplines. For years, data was used in other areas where models were less developed or remained quite inaccurate. Data collected massively was successfully classified, cured, distilled, ... using artificial intelligence (AI) techniques. Thus, correlations between data ca be removed, proving that a certain simplicity remains hidden behind a rather apparent complexity. Data-driven modeling developed exponentially and advanced artificial intelligence techniques were developed, covering six major domains: (i) Multidimensional data visualization [9]; (ii) Data classification and clustering [10,11]; (iii) Learning models from input/output pairs of data, with adequate techniques enabling real-time learning and able to operate in the low-data limit (e.g., sPGD [12], *Code2Vect* [13], iDMD [14–16], NN [17], ThemodynML [5,18], ...); (iv) Knowledge extraction in order to identifying combined parameters and model richness/complexity, discovering hidden parameters, discarding useless parameters or even to extract governing equations; (v) Explaining for certifying; and (vi) Hybridizing physics and data for defining advanced and powerful Dynamic Data-Driven Application Systems, DDDAS [19].

However, these data-driven models, when used in engineering and industry, were quickly confronted with three major and recurrent difficulties: (i) the need for a huge amount of data to make predictions accurate and reliable, knowing that data is synonymous with cost (acquisition and processing costs); (ii) the difficulty of explaining and interpreting predictions obtained by artificial intelligence; and (iii) related to the the latter, the difficulty of certifying engineering products.

1.3. Towards Real-Time Decision Making

In summary, on one side models based on physics can be solved fast but, as discussed, in many engineering areas they remain poor approximations of the real components and systems. On the other hand, when approaching the problem from the data perspective, impressive amounts of data are sometimes needed (with the associated cost and technological difficulty of collecting them), to be processed in real time and then explained in order to certify both the designs and the decisions.

A possible winning option consists of merging both concepts and methodologies. The *hybrid paradigm* was born [19,20], associating in it two type of models: the first based on physics; the second being a completely new type of model, more pragmatic and phenomenological, based on data.

Real-time decision making in engineering design, manufacturing and predictive and operational maintenance, needs the evaluation of quantities of interest in almost real-time. The present work aims at proposing a technique able to determine under the stringent real-time constraints, effective properties of a complex microstructure by assimilating an image of it.

To conciliate accuracy and real-time constraints, the hybrid paradigm is retained: (i) the prediction engine will be trained offline from data coming form physics; then (ii) a non-linear regression, acting on some topological descriptors extracted from those images, will ensure a real-time evaluation of the effective properties (in the present case the homogenized thermal conductivity).

As previously discussed, dealing with shapes and topologies, the parametric description requires performant techniques able to express them in a compact and concise way. In this paper, we propose using Topological Data Analysis, TDA [21], for representing these rich topologies and morphologies, and then using the associated topological descriptors for generating accurate supervised classification and nonlinear regressions, enabling an almost real-time evaluation of the quantities of interest.

In the next section we will present the main methodologies used in the present study, that will be considered later for the training and then for the real-time evaluation of effective properties (homogenized thermal conductivity) of rich microstructures.

2. Methods

This section revisits the main methodologies that will be considered later for real-time classification and prediction of the effective thermal properties from collected images. For that purpose we will consider a rich enough training stage that consists of generating several microstructures whose effective thermal conductivity will be evaluated by using a standard linear homogenization technique, revisited in Section 2.1.

In order to associate the resulting homogenized conductivity tensor to each microstructure, the last must be described in a compact and concise way. For that purpose Topological Data Analysis and Principal Component Analysis (PCA) will be employed. Both are revisited in Sections 2.2 and 2.3, respectively.

The last step aims at performing a nonlinear regression to link the parameters extracted by the TDA to the thermal conductivity. The technique retained in our study is the so-called *Code2Vect* nonlinear regression, revisited in Section 2.4.

2.1. Linear Homogenization Procedure

Due to the microscopic nature of heterogeneity, a procedure is required for extracting the effective thermal conductivity. In what follows we proceed in the linear case, as was also the case in [22], in a representative volume element Ω with a microstructure perfectly defined at that scale. The microscopic conductivity $\mathbf{k}(\mathbf{x})$ is known at every point $\mathbf{x} \in \Omega$.

The macroscopic temperature gradient $\mathbf{G}$ is defined from the space average

$$\mathbf{G} = \langle \mathbf{g}(\mathbf{x}) \rangle \equiv \frac{1}{|\Omega|} \int_\Omega \mathbf{g}(\mathbf{x}) \, d\mathbf{x}, \tag{5}$$

where $\mathbf{g}(\mathbf{x})$ represents the microscopic temperature gradient, i.e., $\mathbf{g}(\mathbf{x}) = \nabla T(\mathbf{x})$.

We define the localization tensor $\mathbf{L}(\mathbf{x})$ such that

$$\mathbf{g}(\mathbf{x}) = \mathbf{L}(\mathbf{x}) \, \mathbf{G}. \tag{6}$$

The microscopic heat flux $\mathbf{q}(\mathbf{x})$ follows the Fourier law

$$\mathbf{q}(\mathbf{x}) = -\mathbf{k}(\mathbf{x}) \, \mathbf{g}(\mathbf{x}), \tag{7}$$

and its macroscopic counterpart $\mathbf{Q}$ reads

$$\mathbf{Q} = \langle \mathbf{q}(\mathbf{x}) \rangle = \langle -\mathbf{k}(\mathbf{x}) \, \mathbf{g}(\mathbf{x}) \rangle = \langle -\mathbf{k}(\mathbf{x}) \, \mathbf{L}(\mathbf{x}) \rangle \, \mathbf{G}, \tag{8}$$

from which the homogenized thermal conductivity reads

$$\mathbf{K} = \langle -\mathbf{k}(\mathbf{x}) \, \mathbf{L}(\mathbf{x}) \rangle. \tag{9}$$

Thus, the calculation of the homogenized thermal conductivity tensor only needs the computation of the tensor $\mathbf{L}(\mathbf{x})$. The present work considers the simplest procedure that in the 2D case consists of solving two steady state thermal problems in Ω

$$\begin{cases} \nabla \cdot (\mathbf{k}(\mathbf{x}) \, \nabla T^1(\mathbf{x})) = 0 \\ T^1(\mathbf{x} \in \partial\Omega) = x, \end{cases} \tag{10}$$

and

$$\begin{cases} \nabla \cdot (\mathbf{k}(\mathbf{x}) \, \nabla T^2(\mathbf{x})) = 0 \\ T^2(\mathbf{x} \in \partial\Omega) = y \end{cases} \tag{11}$$

whose solutions verify by construction

$$\begin{cases} \mathbf{G}^1 = \langle \nabla T^1(\mathbf{x}) \rangle^T = (1,0) \\ \mathbf{G}^2 = \langle \nabla T^2(\mathbf{x}) \rangle^T = (0,1) \end{cases}, \tag{12}$$

and whose gradients define the localization tensor columns

$$\mathbf{L}(\mathbf{x}) = \left(\nabla T^1(\mathbf{x}) \quad \nabla T^2(\mathbf{x}) \right), \tag{13}$$

that allows calculating the effective thermal conductivity.

Remark 1. *In the present work, since we are only interested in the effective thermal conduction along the y-direction, a single problem, problem (11), suffices for calculating the only component of interest, component K_{22}.*

2.2. Topological Data Analysis

Topological data analysis, TDA [21], is one of the most promising techniques in high-dimensional data analysis. In essence, TDA is a powerful tool to find the topology of data: if there are clusters, a manifold structure or even noise that is not relevant for the analysis.

For an intuitive description of the method consider the set of points depicted in Figure 1. In general, these points will live in high dimensional spaces, such that their intrinsic topology will not be visible at first glance. We then equip the set with a distance parameter r. By making r grow, different k-simplexes will appear. Remember that a 0-simplex is a point, a 1-simplex is an edge, a 2-simplex is a triangle, on so on.

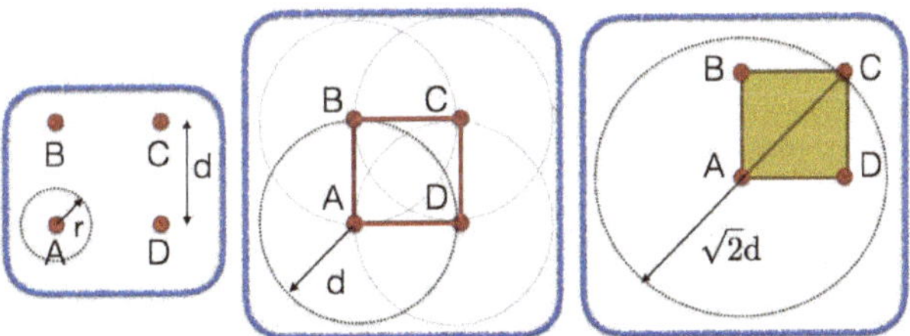

Figure 1. Illustrating TDA: (**left**) For $r < d$ the four points (A, B, C and D) remain disconnected; (**center**) At $r = d$ the hole ABCD appear from the four edges AB, BC, CD and DA; (**right**) The just created hole persist until $r = \sqrt{2}d$, value at which A connects with C and the two resulting triangles ABC and ACD cover the initial hole that disappears consequently.

As r grows, holes appear (as the one defined by the edges between points A, B, C and D in Figure 1, for instance when $r = d$), and disappear for higher values of r (when $r = \sqrt{2}d$, the initial hole is covered by triangles ABC and ACD). Which is important in this discussion is that the overall structure of data is the one that *persists* for longer r values. Holes defined by noisy data are rapidly eliminated from the simple complex.

The value of r at which a hole appears, and then the one at which it disappears, defines a bar joining both, which characterizes the hole persistence. When collecting all the bars associated with all the holes appearing and then disappearing when r grows, the so-called persistence barcode results, the last representing compactly a given morphology.

An alternative consists of using a 2D representation, the so-called persistence diagram (PD), reporting in the x_1-axis the value of r at which a hole appears, and on the x_2-axis the value at which it disappears. Obviously, with the hole birth preceding its death, all the point are place on the upper domain defined by the bisector $x_2 = x_1$, and any point (x_1, x_2) remaining close to that bisector represents noise, a small scale, with the associated hole death following immediately its birth. Points far from the bisector represent the topology that persists.

The persistence barcode and the persistence diagram are two representations with a high physical content; however both representations can not be used for comparison purposes, because they are defined in a non-metric space where the calculation of distances for concluding on proximity has not sense.

To move to a more appropriate space making possible the calculation of distances, we first transform the persistence diagram according to $(x_1, x_2) \rightarrow (y_1 = x_1, y_2 = x_2 - x_1)$ and then apply on the last a convolution (usually with a Gaussian kernel) leading to the so-called persistence image (PI) $\mathbf{y}$, the last defined in a vector space, $\mathbf{y} \in \mathbb{R}^D$, that allows applying most of AI algorithms [23].

2.3. Principal Component Analysis

TDA is able to analyze a complex microstructure through its associated image, and to extract its relevant topological features in form of a persistence image, that can be viewed a matrix. However, using these matrix components is not the most compact and concise way of representing the microstructure, because it contains too many components that makes difficult using it for constructing regressions. Thus, in practice, a linear dimensionality reduction such as principal component analysis (PCA) can be applied for extracting the most representative modes of the persistence images and then to represent in a compact and concise way the microstructures by using the weight associated with the most important modes extracted.

Let us consider a vector $\mathbf{y} \in \mathbb{R}^D$ containing the different components of a persistence image. When considering a set of P microstructures, the associated PIs lead to $\mathbf{y}_i$, $i = 1, \ldots, P$. If they are somehow correlated, there will be a linear transformation $\mathbf{W}$ defining the vector $\boldsymbol{\xi} \in \mathbb{R}^d$, with $d < D$, which contains the still unknown *latent variables*, such that [4]

$$\mathbf{y} = \mathbf{W}\boldsymbol{\xi}. \tag{14}$$

The transformation matrix $\mathbf{W}$, $D \times d$, satisfies the orthogonality condition $\mathbf{W}^T\mathbf{W} = \mathbf{I}_d$, where $\mathbf{I}_d$ represents the $d \times d$ identity matrix.

PCA proceeds by guaranteeing maximal preserved variance and de-correlation in the latent variable set $\boldsymbol{\xi}$. Thus, the covariance matrix of $\boldsymbol{\xi}$,

$$\mathbf{C}_{\xi\xi} = \mathrm{E}\{\boldsymbol{\Xi}\boldsymbol{\Xi}^T\}, \tag{15}$$

will be diagonal. PCA will then extract the d uncorrelated latent variables from

$$\mathbf{C}_{yy} = \mathrm{E}\{\mathbf{Y}\mathbf{Y}^T\} = \mathrm{E}\{\mathbf{W}\boldsymbol{\Xi}\boldsymbol{\Xi}^T\mathbf{W}^T\} = \mathbf{W}\mathrm{E}\{\boldsymbol{\Xi}\boldsymbol{\Xi}^T\}\mathbf{W}^T = \mathbf{W}\mathbf{C}_{\xi\xi}\mathbf{W}^T, \tag{16}$$

that pre- and post-multiplying by $\mathbf{W}^T$ and $\mathbf{W}$, respectively, reads

$$\mathbf{C}_{\xi\xi} = \mathbf{W}^T\mathbf{C}_{yy}\mathbf{W}. \tag{17}$$

By factorizing the covariance matrix $\mathbf{C}_{yy}$, applying the singular value decomposition, SVD,

$$\mathbf{C}_{yy} = \mathbf{V}\boldsymbol{\Lambda}\mathbf{V}^T, \tag{18}$$

and taking into account Equation (17), it results

$$\mathbf{C}_{\xi\xi} = \mathbf{W}^T\mathbf{V}\boldsymbol{\Lambda}\mathbf{V}^T\mathbf{W}, \tag{19}$$

that holds when the d columns of $\mathbf{W}$ are taken collinear with d columns of $\mathbf{V}$, i.e.,

$$\mathbf{W} = \mathbf{V}\mathbf{I}_{D \times d}. \tag{20}$$

2.4. Code2Vect

Code2Vect [13] maps data into a vector space where the distance between points is proportional to the difference of the QoI associated with those points, as sketched in Figure 2.

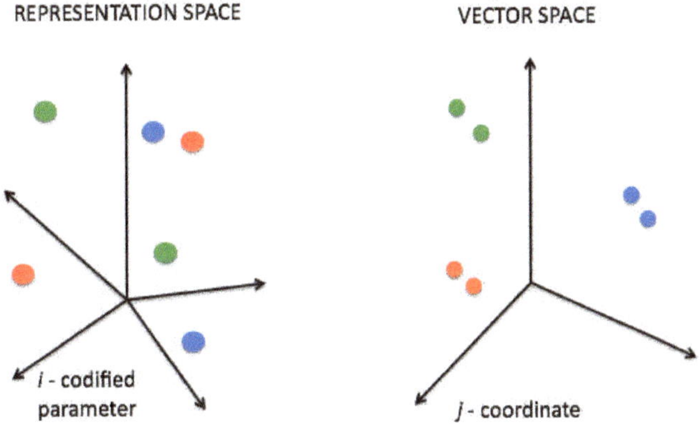

Figure 2. Input space ξ (**left**) and target vector space $\mathbf{z}$ (**right**).

We assume the available data consisting of P d-dimensional arrays, $\xi_i \in \mathbb{R}^d$, with a QoI $\mathcal{O}_i$ associated with each datum. The images, $\mathbf{z}_i \in \mathbb{R}^q$ ($q = 2$ in our numerical implementation for the sake of visualization clarity), results from

$$\mathbf{z}_i = \mathbf{W}\xi_i, \quad i = 1,\ldots,P, \tag{21}$$

that preserves the quantity of interest associated with is origin point ξ_i, denoted by $\mathcal{O}_i$.

In order to place points such that distances scales with their QoI differences we enforce

$$(\mathbf{W}(\xi_i - \xi_j)) \cdot \mathbf{W}(\xi_i - \xi_j)) = \|\mathbf{z}_i - \mathbf{z}_j\|^2 = |\mathcal{O}_i - \mathcal{O}_j|. \tag{22}$$

Thus, there are $\frac{P^2}{2} - P$ relations to determine the $q \times d + P \times q$ unknowns. Linear mappings are limited and do not allow proceeding in nonlinear settings. Thus, a better choice consists of a nonlinear mapping $\mathbf{W}(\xi)$, expressible as a general polynomial form.

3. Results

3.1. Model Training

Several microstructures, based on a population of holes (from now on called pores) with different sizes, shapes, location, and number of pores, distributed in the 2D square domain Ω, are created. Four of these microstructures are shown in Figure 3. They are equipped with a mesh on which finite element calculations will be done for computing the reference effective (homogenized) thermal conductivity, in particular the component K_{22} of the homogenized conductivity tensor.

These meshes also serve to apply the TDA in order to obtain the persistence diagram (PD) and its associated persistence image (PI). As previously indicated, the last consists of a convolution applied on the former. Each persistence image defines a 20×20 matrix, or its vector counterpart $\mathbf{y}_i \in \mathbb{R}^{400}$.

Thus, TDA is able to analyze a complex microstructure through its image, and extract its relevant topological features in form of a persistence image, that can be viewed as a matrix. However, this matrix still contains too much information (its number of components, here 20×20) to perform classification or regression when not too much data is available (scarce-data limit). Obviously, large amounts of synthetic data can be produced by solving numerically thousands or even millions of thermal problems. However, in engineering cheap solutions are usually preferred, and in particular smart-data is preferred to its big counterpart. Efficiency seems a better option that brute force, and for this reason, here we prefer keeping the amount of data as reduced as possible, and compensate its absence by enhancing the amount of information that data contains.

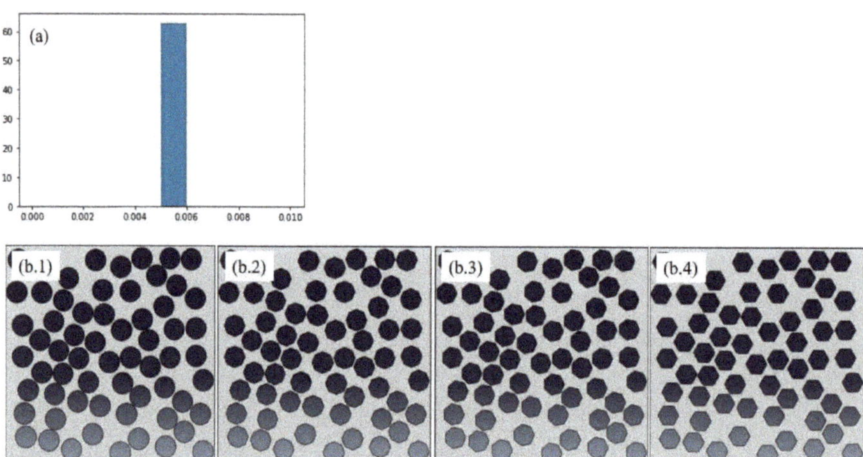

Figure 3. (**a**) Histogram of pores radius; (**b**) Pore shapes: Circle, Octagon, Heptagon and Hexagon.

Thus, persistence images are still not the most compact way of representing the topological and morphological features of the analyzed microstructures. For improving the representation we apply a linear dimensionality reduction, the principal component analysis, for extracting the most representative modes of the persistence images. Thus, the weights of those PCA modes will constitute the compact and concise way to represent those microstructures.

From a practical viewpoint PCA allowed reducing from $400 = (20 \times 20)$ the dimension of PI resulting from TDA, to 3 dimensions. Thus, each analyzed microstructure is concisely represented by 3 coordinates (the weights of the first three most relevant PCA modes) and each one has attached a QoI, the effective thermal conductivity K_{22} obtained from a finite element simulation following the rationale described in Section 2.1. Now, the nonlinear regression relating the output, the QoI (the effective thermal conductivity in our case), with the parameters describing the microstructure, the three PCA weights, is performed by applying the *Code2Vect* nonlinear regression, summarized in Section 2.4.

As soon as the regression is constructed at the present training stage, it could be used online for predicting the conductivity of new microstructures.

3.2. Inferring Effective Properties

We prepared 5 samples, four of them were used in the training stage, represented in Figure 3, in which the pores volume fraction was kept constant ($\phi = 0.5$) and the spatial distribution almost uniform.

The constructed nonlinear regression (based on the use of *Code2Vect*) described in the previous section, is now applied to the sample shown in Figure 4 where while keeping the same almost uniform pore distribution and the same volume fraction, hexagons and heptagons were randomly mixed. In this same figure, the solution of the thermal problem at the microscopic scale for obtaining the effective thermal conductivity that will serve as reference value, is also included. Finally, it also shows both the PD and the PI.

The PI, $\mathbf{y} \in \mathbb{R}^D$, is then projected into the three retained orthonormal PCA modes to give the thee weights that constitute the data $\boldsymbol{\xi} \in \mathbb{R}^3$ ($d = 3$) to be processed by the nonlinear regression (based on the *Code2Vect*) that produces vector $\mathbf{z} \in \mathbb{R}^2$ (we enforce a 2D representation, $q = 2$, for the sake of clarity in the data visualization)

$$\mathbf{z} = \mathbf{W}(\xi)\,\xi, \tag{23}$$

and then identify the set $\mathcal{S}(\mathbf{z})$ of data $\mathbf{z}_i$ closest to $\mathbf{z}$, from which the QoI, the effective thermal conductivity, is interpolated

$$\mathcal{O} = \sum_{i \in \mathcal{S}(\mathbf{z})} \mathcal{F}(\mathbf{z}, \mathbf{z}_i)\,\mathcal{O}_i, \tag{24}$$

with in the present case $\mathcal{O} \equiv K_{22}$ and with radial bases as interpolation functions $\mathcal{F}(\mathbf{z}, \mathbf{z}_i)$.

Figure 5 places $\mathbf{z}$ with respect to its neighbors, where color scales with the target quantity, that is, with K_{22}. The inferred value of the effective thermal conductivity K_{22} using Equation (24) for the micorstructure depcited in Figure 4 results $K_{22}(\mathbf{z}) = 73.4$ W/mK, very close to the reference value computed numerically from the temperature distribution shown also in Figure 4, of $K_{22,\mathrm{REF}} = 74$ W/mK.

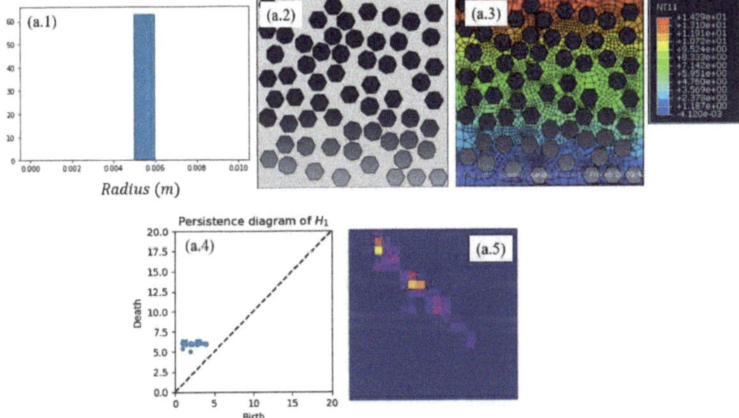

Figure 4. (**a.1**) Histogram of the pores radius; (**a.2**) considered microstructure; (**a.3**) temperature field used for computing the effective thermal conductivity that will serve as reference for evaluating the regression performance; (**a.4**) persistence diagram; and (**a.5**) persistence image.

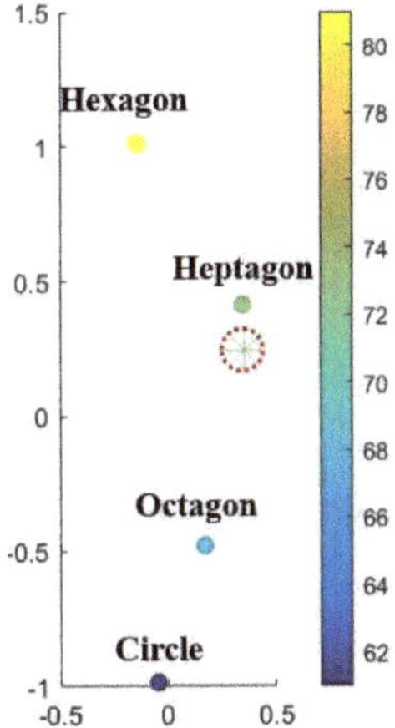

Figure 5. Interpolation space $\mathbf{z}$ with color scaling with the values of the effective thermal conductivity K_{22}.

3.3. Microstructures with Varying Shapes and Size Distribution

These first preliminary successful results were pushed forward by considering quite more complex microstructures. Thus, a total of 35 samples were generated, while varying other parameters, in particular pore size (following uniform and Gamma distributions) and pore shape (circular or 5 to 8 side polygons, randomly chosen). The volume fraction was kept constant ($\phi = 0.5$). 34 samples were used in the training, keeping one, the one shown in Figure 6, for inferring the thermal conductivity and concluding on the ability of the proposed technique to infer accurately it. Figure 7 places the considered microstructure in the **z**-space where the thermal conductivity is interpolated, to infer the value of $K_{22} = 81$ W/mK, for a reference value of $K_{22,\text{REF}} = 78$ W/mK.

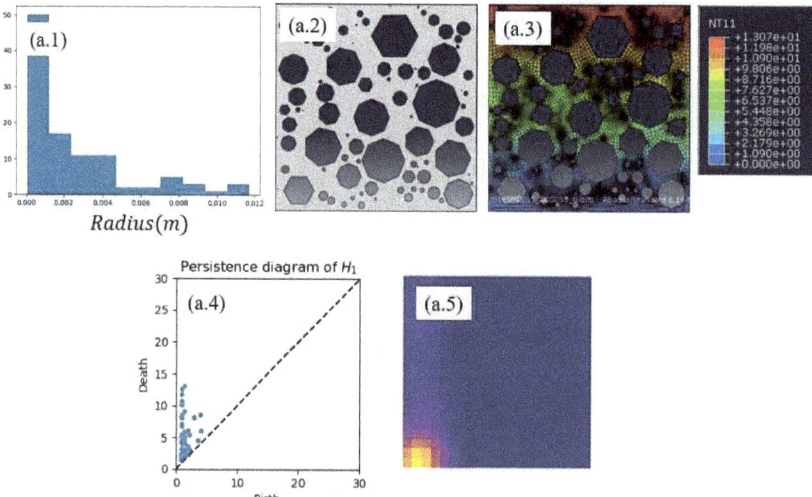

Figure 6. (**a.1**) Histogram of pores radius; (**a.2**) testing microstructure; (**a.3**) temperature field used for calculating the reference effective thermal conductivity; (**a.4**) persistence diagram; and (**a.5**) persistence image.

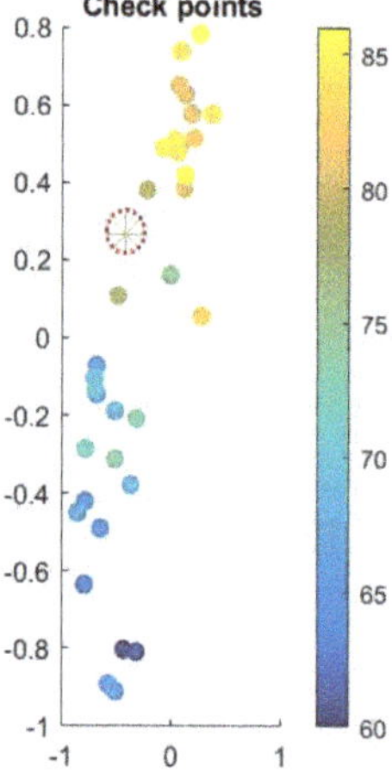

Figure 7. Interpolation space **z** with color scaling with the values of the effective thermal conductivity K_{22}.

To check the prediction improvement with the sampling richness, the effective thermal conductivity in the microstructure shown in Figure 6 while considering different samplings in the training stage, from 13 to 35 microstructures, with the relative errors reported in Table 1.

Table 1. Relative error in the effective conductivity prediction depending on the number of samples considered in the regression (training stage).

Number of Samples	Relative Error
13	0.076
16	0.056
19	0.046
35	0.037

4. Conclusions

The present study proves that effective properties can be associated with microstructures with complex morphological and topological features. For this purpose, those features are extracted by using TDA, post-compressed by using linear dimensionality reduction (PCA) which output represents the parameters employed by the nonlinear *Code2Vect* regression that finally assign a effective property (here the effective thermal conductivity) to a given microstructure.

The procedure demonstrated its robustness and performance in the low-data limit, as well as its capacity to provide better predictions when considering larger training sets. It successfully combines physics-based data for learning purposes, with almost real-time inference based on the topological analysis of images.

Author Contributions: Conceptualization, software and validation, M.Y. and C.A.; methodology, J.L.D., E.C. and F.C. All authors have read and agreed to the published version of the manuscript.

Funding: This research received no external funding.

Acknowledgments: The authors acknowledge the ESI Chairs at Arts et Métiers Institute of Technology and the University of Zaragoza, as well as the French ANR through the DataBEST project.

Conflicts of Interest: The authors declare no conflict of interest.

References

1. Chinesta, F.; Huerta, A.; Rozza, G.; Willcox, K. *Model Order Reduction Chapter in the Encyclopedia of Computational Mechanics*, 2nd ed.; John Wiley & Sons, Ltd.: Hoboken, NJ, USA, 2015.
2. Chinesta, F.; Leygue, A.; Bordeu, F.; Aguado, J.V.; Cueto, E.; González, D.; Alfaro, I.; Ammar, A.; Huerta, A. Parametric PGD based computational vademecum for efficient design, optimization and control. *Arch. Comput. Methods Eng.* **2013**, *20*, 31–59. [CrossRef]
3. Chinesta, F.; Keunings, R.; Leygue, A. *The Proper Generalized Decomposition for Advanced Numerical Simulations*; Springerbriefs, Springer: Berlin, Germany, 2014.
4. Lee, J.A.; Verleysen, M. *Nonlinear Dimensionality Reduction*; Springer: New York, NY, USA, 2007.
5. González, D.; Chinesta, F.; Cueto, E. Thermodynamically consistent data-driven computational mechanics. Continuum Mech. *Thermodynamics* **2018**, *31*, 239–253. [CrossRef]
6. González, D.; Cueto, E.; Chinesta, F. Computational patient avatars for surgery planning. *Ann. Biomed. Eng.* **2016**, *44*, 35–45. [CrossRef] [PubMed]
7. Lopez, E.; Gonzalez, D.; Aguado, J.V.; Abisset-Chavanne, E.; Cueto, E.; Binetruy, C.; Chinesta, F. A manifold learning approach for integrated computational materials engineering. *Arch. Comput. Methods Eng.* **2016**, *25*, 59–68. [CrossRef]
8. Gonzalez, D.; Aguado, J.V.; Cueto, E.; Abisset-Chavanne, E.; Chinesta, F. kPCA-based Parametric Solutions within the PGD Framework. *Arch. Comput. Methods Eng.* **2018**, *25*, 69–86. [CrossRef]
9. Maaten, L.; Hinton, G. Visualizing data using t-SNE. *J. Mach. Learn. Res.* **2008**, *9*, 2579–2605.

10. Cristianini, N.; Shawe-Taylor, J. *An Introduction to Support Vector Machines: And other Kernel-Based Learning Methods*; Cambridge University Press: New York, NY, USA, 2000.

11. Forgy, E. Cluster analysis of multivariate data: Efficiency versus interpretability of classification. *Biometrics* **1965**, *21*, 768–769.

12. Ibanez, R.; Abisset-Chavanne, E.; Ammar, A.; Gonzalez, D.; Cueto, E.; Huerta, A.; Duval, J.L.; Chinesta, F. A multi-dimensional data-driven sparse identification technique: The sparse Proper Generalized Decomposition. *Complexity* **2018**, 5608286. [CrossRef]

13. Argerich, C.; Ibanez, R.; Barasinski, A.; Chinesta, F. Code2vect: An efficient heterogenous data classifier and nonlinear regression technique. *Comptes Rendus Mécanique* **2019**, *347*, 754–761. [CrossRef]

14. Reille, A.; Hascoet, N.; Ghnatios, C.; Ammar, A.; Cueto, E.; Duval, J.L.; Chinesta, F.; Keunings, R. Incremental dynamic mode decomposition: A reduced-model learner operating at the low-data limit. *Comptes Rendus Mécanique* **2019**, *347*, 780–792. [CrossRef]

15. Schmid, P.J. Dynamic mode decomposition of numerical and experimental data. *J. Fluid Mech.* **2010**, *656*, 5–28. [CrossRef]

16. Williams, M.O.; Kevrekidis, G.; Rowley, C.W. A data-driven approximation of the Koopman operator: Extending dynamic mode decomposition. *J. Nonlinear Sci.* **2015**, *25*, 1307–1346. [CrossRef]

17. Goodfellow, I.; Bengio, Y.; Courville, A. *Deep Learning*; MIT Press: Cambridge, UK, 2016

18. Moya, B.; Gonzalez, D.; Alfaro, I.; Chinesta, F.; Cueto, E. Learning slosh dynamics by means of data. *Comput. Mech.* **2019**, *64*, 511–523. [CrossRef]

19. Chinesta, F.; Cueto, E.; Abisset-Chavanne, E.; Duval, J.L.; El Khaldi, F. Virtual, Digital and Hybrid Twins: A New Paradigm in Data-Based Engineering and Engineered Data. *Arch. Comput. Methods Eng.* **2020**, *27*, 105–134. [CrossRef]

20. González, D.; Chinesta, F.; Cueto, E. Learning corrections for hyperelastic models from data. *Front. Mater. Comput. Mater. Sci.* **2019**, *6*. [CrossRef]

21. Wasserman, L. Topological data analysis. *Ann. Rev. Stat. Appl.* **2018**, *5*, 501–532. [CrossRef]

22. Lamari, H.; Ammar, A.; Cartraud, P.; Legrain, G.; Jacquemin, F.; Chinesta, F. Routes for Efficient Computational Homogenization of Non-Linear Materials Using the Proper Generalized Decomposition. *Arch. Comput. Methods Eng.* **2010**, *17*, 373–391. [CrossRef]

23. Adams, H.; Emerson, T.; Kirby, M.; Neville, R.; Peterson, C.; Shipman, P.; Chepushtanova, S.; Hanson, E.; Motta, F.; Ziegelmeier, L. Persistence images: A stable vector representation of persistent homology. *J. Mach. Learn. Res.* **2017**, *18*, 218–252.

Article

A Methodology for the Statistical Calibration of Complex Constitutive Material Models: Application to Temperature-Dependent Elasto-Visco-Plastic Materials

Juan Luis de Pablos [1,2], **Edoardo Menga** [3] **and Ignacio Romero** [1,2,*]

1   IMDEA Materials Institute, Eric Kandel, 2, 28906 Getafe, Spain; juanluis.pablos@imdea.org
2   Mechanical Eng. Department, Universidad Politécnica de Madrid, José Gutiérrez Abascal, 2, 28006 Madrid, Spain
3   AIRBUS Operations S.L., John Lennon S/N, 28906 Getafe, Spain; edoardo.menga@airbus.com
*   Correspondence: ignacio.romero@upm.es

Received: 18 August 2020; Accepted: 28 September 2020; Published: 2 October 2020

Abstract: The calibration of any sophisticated model, and in particular a constitutive relation, is a complex problem that has a direct impact in the cost of generating experimental data and the accuracy of its prediction capacity. In this work, we address this common situation using a two-stage procedure. In order to evaluate the sensitivity of the model to its parameters, the first step in our approach consists of formulating a meta-model and employing it to identify the most relevant parameters. In the second step, a Bayesian calibration is performed on the most influential parameters of the model in order to obtain an optimal mean value and its associated uncertainty. We claim that this strategy is very efficient for a wide range of applications and can guide the design of experiments, thus reducing test campaigns and computational costs. Moreover, the use of Gaussian processes together with Bayesian calibration effectively combines the information coming from experiments and numerical simulations. The framework described is applied to the calibration of three widely employed material constitutive relations for metals under high strain rates and temperatures, namely, the Johnson–Cook, Zerilli–Armstrong, and Arrhenius models.

Keywords: model calibration; sensitivity analysis; elasto-visco-plasticity; Gaussian process

---

1. Introduction

Modeling has become a very effective way to analyze, in a first instance, complex engineering problems. Almost the totality of engineers, either in academia or in the industry, claim to take benefit from these techniques, considering them to be irreplaceable for their work.

Though the reliability of models keeps constantly increasing, and therefore the trust placed on their predictions, there is still need for understanding the intrinsic uncertainties that affect simulation, for estimating their effect on predictions, and for developing efficient methodologies to reduce them in a cost-effective manner. In this respect, an interesting and promising approach has emerged in recent years. It consists of employing advanced statistical methods not only to assess the uncertainty in a model but also to guide the experimental campaign that needs to be carried out to feed the parameter calibration. One of these tools is Global Sensitivity Analysis (GSA), a very useful strategy when it comes to analyzing the influence of all the parameters participating in a model. Improving local techniques introduced in the 1980s [1], GSA methods were proposed much later to account for the influence of parameters in an overall and rigorous fashion [2].

One important limitation of GSA techniques is that they require large amounts of simulated data as input. Numerical experiments obtained, for instance, through finite element (FE) simulations, demand a huge amount of computational resources that are often unavailable. A convenient remedy to this problem is to employ meta-models that provide reasonable approximations to the models' response, but at a fraction of their computational cost. These types of models are built by sampling the original ones, as illustrated in Figure 1, and were originally proposed to optimize processes [3]. Initially known as Response Surface Models (RSM), they rapidly evolved and became emulators of computational codes at a very reduced cost. As a result, they have been utilized in a wide range of sensitivity analyses and applications [4–6].

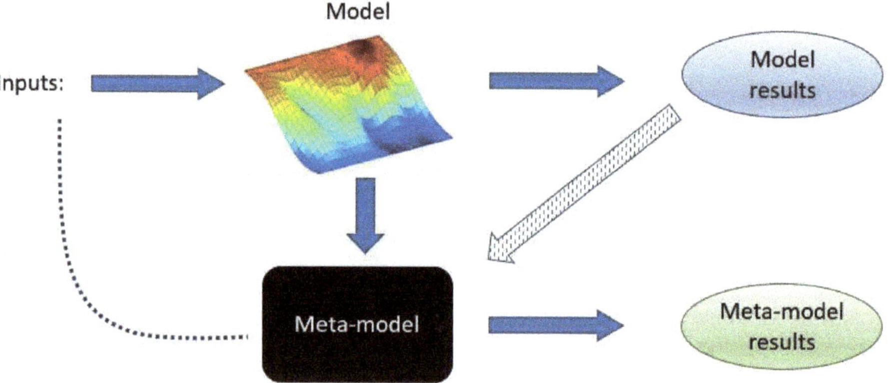

Figure 1. Meta-modeling construction process.

There exist several families of meta-models. Some of the most commonly employed are the ones based on Kriging [7] and Radial Basis Functions (RBF) [8]. Both of them are generally accepted as good methods to efficiently capture trends associated with small data sets. Since they accurately adapt to available information, they must be re-calibrated when new inputs are provided [9]. The Bayesian approach, on the other hand, is a well-known technique that has been successfully employed in several scientific disciplines for parameter selection. See, for example, one of the very first generic applications, developed by Guttman [10], in which this inference procedure is already used to choose the best manufacturing parameters to make the widest possible population of fabricated items lie within the specified tolerance limits. More recent works have improved the Bayesian inference methodology (see, e.g., [11]).

Bayesian inference can be used systematically for the calibration of model parameters, taking into consideration the uncertainties due to the model itself, the experimental measurements, noise, etc. [12–14]. This approach has become relatively standard, not only providing optimized parameter values but a complete Gaussian distribution for them.

In this work, we will combine GSA with Bayesian calibration because we believe that this combination is extremely powerful for understanding computer models, and drawing as much information as possible from experiments, be them numerical or physical. This mix of techniques is not new, and similar ones have been considered in the past. For example, sensitivity analysis and Bayesian calibration were employed together in [15], trying to assess multiple sources of uncertainty in waste disposal models by considering independent and composite scenarios, obtaining predictive output distributions using a Bayesian approach, and later performing a variance-based sensitivity analysis. In addition, the work by [16] proposes a procedure to evaluate the sensitivity of the parameters and a posterior calibration of the most important ones applied to a model describing the chemical composition of mass of waters.

In the current article, we explore the use of GSA and Bayesian calibration for complex constitutive models, an application that has not been previously considered for this type of analysis and that can greatly benefit from it. More precisely, we study three fairly known constitutive material models suitable for metals subjected to extreme conditions, namely, Johnson–Cook [17], Zerilli–Armstrong [18], and Arrhenius-type [19] models. These are fairly complex constitutive relations that depend on a relatively large number of material parameters that need to be adjusted for each specific material and test range. The actual implementations of the three can be found in the publicly available material library MUESLI [20], and we have used them together with standard explicit finite element calculations.

The remainder of the article is structured as follows. In Section 2, we will outline the theoretical principles in which the statistical theory employed is based on, as well as the three constitutive material models used in the study. In Section 3, we describe the application of the presented framework to the analysis of Taylor's impact test [21,22], an experiment often used to characterize the elastoplastic behavior of metals under high strain rates. The results of our investigation are reported in Section 4, providing insights for the three constitutive models. Finally, Section 5 collects the main findings and conclusions of the study.

2. Fundamentals

2.1. Global Sensitivity Analysis

Global Sensitivity Analysis (GSA) refers to a collection of techniques that allow identifying the most relevant variables in a model with respect to given Quantities of Interest (QoIs). They focus on apportioning the output's uncertainty to the different sources of uncertain parameters [2] and define qualitative and quantitative mappings between the multi-dimensional space of stochastic variables in the input and the output. The most popular GSA techniques are based on the decomposition of the variance's output probability distribution and allow the calculation of Sobol's sensitivity indices.

According to Sobol's decomposition theory, a function can be approximated as the sum of functions of increasing dimensionality and orthogonal with respect to the standard L_2 inner product. Hence, given a mathematical model $y = f(x)$ with n parameters arranged in the input vector x, the decomposition can be expressed as:

$$f(x_1, \ldots, x_n) = f_0 + \sum_{i=1}^{n} f_i(x_i) + \sum_{1 \leq i < j \leq n} f_{ij}(x_i, x_j) + \cdots + f_{1,2,\ldots,n}(x_1, \ldots, x_n), \tag{1}$$

where f_0 is a constant and $f_{\ldots}$ are functions with domains of increasing dimensionality. If we consider that f is defined on random variables $X_i \sim \mathcal{U}(0,1), i = 1, \ldots, n$, then the model output is itself a random variable with variance

$$D = \text{Var}[f] = \int_{\mathbb{R}^n} f^2(x)\, dx - f_o^2. \tag{2}$$

Integrating Equation (1) and using the orthogonality property of the functions $f_{\ldots}$, we note that the variance itself can be decomposed in the sum

$$D = \sum_{i=1}^{n} D_i + \sum_{1 \leq i < j \leq n} D_{ij} + \cdots + D_{1,2,\ldots,n}. \tag{3}$$

This expression motivates the definition of the Sobol indices

$$S_{i_1,\ldots,i_s} = D_{i_1,\ldots,i_s} / D, \tag{4}$$

that trivially satisfy

$$\sum_{i=1}^{n} S_i + \sum_{1 \le i < j \le n} S_{ij} + \cdots + S_{1,2,\ldots,n} = 1. \tag{5}$$

This decomposition of the total variance D reveals the main metrics employed to assess the relevance of each parameter in the scatter of the quantity of interest f. The relative variances S_i are referred to as the first order indices or main effects and gauge the influence of the i-th parameter on the model's output. The total effect or total order sensitivities associated with the $i-$parameter, including its influence when combined with other parameters, is calculated as

$$S_{T_i} = \sum_{\mathcal{I}_i} D_{i_1,\ldots,i_s}, \ \mathcal{I}_i = \{(i_1,\ldots,i_s) : \exists k, 1 \le k \le s, i_k = i\}. \tag{6}$$

The widespread use of the main and total effects as sensitivity measures is due to the relative simplicity of the formulas and algorithms that can be employed to calculate or approximate them ([2], Chapter 4). Specifically, the number of simulations required to evaluate these measures is $N_s(n+2)$, where N_s is the so-called base sample, a number that depends on the model complexity varying from a few hundreds to several thousands, and n is, as before, the number of parameters of the model (we refer to ([2], Chapter 4) for details on these figures). To calculate sensitivity indices, it proves essential to first create a simple meta-model that approximates the true model, because a limited set of runs is supposed to suffice for building a surrogate that, demanding far fewer computational resources, can then be run a large number of times to complete the GSA.

2.2. Meta-Models

A meta-model is a model for another model. That is, a much-simplified version of a given model that can provide, however, similar predictions as the original one for the same values of the parameters. In this work, we restrict our study to linear meta-models. To describe them, let us assume that the model we are trying to simplify depends on N_p parameters and we denote the quantity of interest, assumed for simplicity to be a scalar, as y. For any collection of parameters $x \in \mathbb{R}^{N_p}$, we define $\hat{y}(x)$ to be the corresponding value of the quantity of interest and slightly abusing the notation we write

$$\hat{y}(x) = [\hat{y}(x_1), \hat{y}(x_2), \ldots, \hat{y}(x_N)], \tag{7}$$

where N is the number of samples and $x = [x_1, x_2, \ldots, x_N]$ is an array of N samples of the parameters. Then, a meta-model is a function $\hat{Y} : \mathbb{R}^{N_p} \to \mathbb{R}$ that approximates $\hat{y}$ and is of the form

$$\hat{Y}(x) := \sum_{k=1}^{N_k} \eta_k h_k(x). \tag{8}$$

In this equation, and later, h_k are the kernels of the approximation and η_k refer to the weights. Abusing slightly the notation again, we express the relation between the model and its meta-model as

$$\hat{y}(x) \approx \hat{Y}(x) = \sum_{k=1}^{N_k} \eta_k h_k(x), \tag{9}$$

or in compact form,

$$\hat{y}(x) \approx \hat{Y}(x) := H(x)\eta, \tag{10}$$

where H is the so-called kernel matrix that collects all the kernel functions.

The precise definition of a meta-model depends, hence, on the number and type of kernel functions h_k and the value of the regression coefficients η_k. Given an a priori choice for the kernels, the weights can be obtained from a set of model evaluations $\hat{y}(x)$ employing a least-squares minimization. Given, as before, an array of sample parameters x and their model evaluation $\hat{y}(x)$, the vector η can be calculated in closed form with the solution of the normal equations to the approximation. That is,

$$\eta = \left(H(x)^T H(x)\right)^{-1} H(x)^T \hat{y}(x). \tag{11}$$

As previously indicated, the kernel functions belong to a set that must be selected a priori. In the literature, several classes of kernel have been proposed for approximating purposes and in this work we select anisotropic Radial Basis Functions (RBF). This type of kernel has shown improved accuracy as compared with standard RBF, particularly when only a limited set of model evaluations is available [23].

A standard RBF is a map $K : \mathbb{R}^{N_p} \times \mathbb{R}^{N_p} \to \mathbb{R}$ of the form

$$K(x,z) = k(r(x,z)), \tag{12}$$

where $k : \mathbb{R} \to \mathbb{R}$ is a monotonically decreasing function and $r(x,z) := |x - z|$ is the Euclidean distance. The anisotropic radial kernels redefine the function K to be of the form

$$K(x,z) = \exp[-\epsilon \sum_{i=1}^{N_d} \gamma_i^2 (x_i - z_i)^2] = \exp[-\epsilon(x-z)^T \Gamma (x-z)], \tag{13}$$

where $\Gamma = \mathrm{diag}(\gamma_1^2, \ldots, \gamma_{N_d}^2)$ is a diagonal, positive definite matrix that scales anisotropically the contribution of each direction in the difference $x - z$, and $\epsilon > 0$ is the shape parameter of the kernel.

2.3. Bayesian Inference and Gaussian Processes

Bayesian inference is a mathematical technique used to improve the knowledge of probability distributions extracting information from sampled data [24]. It is successfully employed for a wide variety of applications in data science and applied sciences. For instance, it has been used in conjunction with techniques such as machine learning and deep learning in fields like medicine [25], robotics [26], earth sciences [27], and more. In this article, we employ Bayesian methods to find the optimal value of model parameters as a function of prior information and observed data. More specifically, we are concerned with model calibration and using it in combination with meta-models to obtain realistic parameter values for complex constitutive relations and their uncertainty quantification, with an affordable computational cost.

Some of the most robust techniques for calibration are based on non-parametric models for nonlinear regression [28]. Here, we will employ Gaussian processes to represent in an abstract fashion the response of a simulation code to a complex mechanical problem employing the material models that we are set to study. We summarize next the main concepts behind these processes.

A Gaussian process is a set of random variables such that any finite subset of them has a Gaussian multivariate distribution [28]. Such a process is completely defined by its mean and covariance, which are functions. If the set of random variables is indexed by points $x \in \mathcal{X} \subset \mathbb{R}^d$, then when the random variables have a scalar value $f(x)$, the standard notation employed is

$$f(x) \sim \mathcal{GP}\left(m(x), k(x, x')\right), \tag{14}$$

where $m : \mathcal{X} \to \mathbb{R}$ and $k : \mathcal{X} \times \mathcal{X} \to \mathbb{R}$ are, respectively, the mean and covariance. The mean function can be arbitrary, but the covariance must be a positive function. Often these two functions are given explicit expressions depending on hyperparameters. In simple cases, the average is assumed to be zero, but often it is assumed to be of the form

$$m(x) = g(x)^T \beta, \tag{15}$$

where $g : \mathbb{R}^d \to \mathbb{R}^g$ is a vector of known basis functions and $\beta \in \mathbb{R}^g$ is a vector of basis coefficients. The choice of the covariance function is the key aspect that determines the properties of the Gaussian process. It is often selected to be stationary, that is, depending only on a distance $d = \hat{d}(x, x')$. In particular, we will employ a covariance function such as

$$c(x, x') = \sigma^2 r(x, x'), \tag{16}$$

where σ^2 is a variance hyperparameter and $r : \mathbb{R}^d \times \mathbb{R}^d \to \mathbb{R}$ has been chosen as the Màtern $C_{5/2}$ function, an isotropic, differentiable, stationary kernel commonly used in statistical fitting that is of the form

$$r(x, x') = \left(1 + \sqrt{5}\frac{\hat{d}^2(x, x')}{\psi} + 5\frac{\hat{d}^2(x, x')}{3\psi^2}\right) \exp\left[-\sqrt{3}\frac{\hat{d}^2(x, x')}{\psi}\right], \tag{17}$$

that uses a length-scale hyperparameter ψ. For the Gaussian process described, the collection of hyperparameters can be collected as $\chi = (\beta, \sigma^2, \psi)$.

Let us now describe in which sense Bayesian analysis can be used for model calibration. The value of a computer model depends on the value of some input variables x that are measurable, and some parameters t that are difficult to determine because they are not directly measurable. Let us assume that $t = \theta$ is the true value of the parameters, which is unknown and we would like to determine based on available data. Given some input variable x, a physical experiment of the problem we want to model will produce a scalar output z and it will verify

$$z = \eta(x, \theta) + \delta(x) + \varepsilon(x) . \tag{18}$$

In this equation, $\eta(x, \theta)$ is the value of the computer model evaluated at the input variable x and the true parameter θ, $\delta(x)$ is the so-called model inadequacy and $\varepsilon(x)$ is the observation error. This last term can be taken to be a random variable with a Gaussian probability distribution $\mathcal{N}(0, \lambda^2)$. The functions η and δ are completely unknown so we can assume them to be Gaussian processes with hyperparameters χ_η and χ_δ, respectively.

If $\theta, \chi_\eta, \chi_\delta, \lambda^2$ were known, we could study the multivariate probability distribution of the output z using Equation (18) for any set of inputs $(x_1, x_2, \ldots, x_s)$. However, we are interested in solving the inverse problem: we have a set of experimental and computational data and we would like to determine the most likely probability distribution for θ and the hyperparameters, a problem that can be effectively addressed using Bayes' theorem.

Bayes' theorem states that, given a prior probability for the parameters $(\theta, \chi_\eta, \chi_\delta, \lambda^2)$ indicated as $p(\theta, \chi_\eta, \chi_\delta, \lambda^2)$, the posterior probability density function for these parameters after obtaining the data Δ is

$$p(\theta, \chi_\eta, \chi_\delta, \lambda^2 | \Delta) \propto p(\theta, \chi_\eta, \chi_\delta, \lambda^2) \, p(\Delta | \theta, \chi_\eta, \chi_\delta, \lambda^2) . \tag{19}$$

The prior for the parameters and hyperparameters can be taken as Gaussian, or any other probability distribution that fits our initial knowledge. Assuming that the parameters and hyperparameters are independent, we have in any case that

$$p(\theta, \chi_\eta, \chi_\delta, \lambda^2) = p(\theta) \, p(\chi_\eta, \chi_\delta, \lambda^2) . \tag{20}$$

In addition, since Equation (18) indicates that the output is the sum of three random variables with Gaussian distributions, z itself is a Gaussian.

To apply Bayes' theorem, it remains to calculate the likelihood $p(\Delta | \theta, \chi_\eta, \chi_\delta, \lambda^2)$. To this end, the hyperparameters of η and δ are collected together with the observation error ε and we assume that the conditioned random variable $\Delta | \theta, \chi_\eta, \chi_\delta, \lambda^2$ has a normal probability distribution of the form

$$N\{E[\Delta | \theta, \chi_\eta, \chi_\delta, \lambda^2], \mathrm{Var}[\Delta | \theta, \chi_\eta, \chi_\delta, \lambda^2]\}. \tag{21}$$

Here, $E[\cdot]$ and $Var[\cdot]$ refer to the expectation and variance, respectively. Finally, in order to obtain the segregated posterior probability density of the parameters $p(\theta|\Delta)$, we should integrate out χ_η, χ_δ and λ^2; but, due to the high computational cost involved, this is typically done using Monte Carlo methods [29]. Details of this process fall outside the scope of the present work and can be found in standard references [12].

2.4. Material Models

In Section 4, a sensitivity and calibration procedure is applied to three relatively complex material models employed in advanced industrial applications. Here, we summarize them, listing all the parameters involved in their description. The remainder of the article will focus on ranking the significance of these parameters, their influence in the predictions, and the determination of their probability distributions.

2.4.1. Johnson–Cook Constitutive Relations

The Johnson–Cook (JC) constitutive model [17] is commonly used to reproduce the behavior of metals subjected to large strains, high strain rates, and high temperatures. It is not extremely accurate in all ranges of strain rates and temperatures, but it is simple to implement, robust, and, as a result, has been employed over the years in a large number of applications [30–32].

Johnson–Cook's model is a J_2 classical plasticity constitutive law in which the yield stress σ_y is assumed to be of the form:

$$\sigma_y = (A + B\varepsilon_p^n)(1 + C\log\dot{\varepsilon}_p^*)(1 - T^{*m}), \tag{22}$$

where ε_p refers to the equivalent plastic strain. The first term in expression (22) accounts for the quasistatic hardening, including the initial yield stress, A, and the constants representing the strain hardening, B, and the exponent n. The second term in (22) is related to the hardening due to strain rate effects, containing the strain rate constant, C and also the dimensionless plastic strain rate, $\dot{\varepsilon}^* = \frac{\dot{\varepsilon}_p}{\dot{\varepsilon}_{p0}}$, where $\dot{\varepsilon}_{p0}$ is a reference plastic strain rate, often taken to be equal to 1. Finally, the third term accounts for the effects of temperature, including the thermal softening exponent, m, and also the so-called "homologous temperature"

$$T^* = \frac{T_{exp} - T_{room}}{T_{melt} - T_{room}}. \tag{23}$$

Here, T_{exp} is the experimental temperature at which the material is being modeled, T_{melt} is the melting temperature of the material and T_{room} is the ambient temperature.

2.4.2. Zerilli–Armstrong Constitutive Relations

The Zerilli–Armstrong (ZA) model [18] was conceived as a new approach to the metal plasticity modeling, using a finite deformation formulation based on constitutive relations related to the physical phenomenon of dislocation mechanics, in contrast to other purely phenomenological constitutive relations, such as the previously described Johnson–Cook model. These relations have been proved to be well suited for the modeling of the response of metals to high strains, strain rates, and temperatures. Its numerical implementation, although more complicated than the JC model, is still relatively simple, justifying its popularity.

The ZA relations were developed in order to respond to the need of a physical-based model that could include the high dependence of the flow stress of metals and alloys on dislocation mechanics. For instance, aspects like the grain size, thermal activation energy, or the characteristic crystal unit cell structure have a dramatic effect in the plastic response of these materials, according to experimental data. Hence, the ZA model is still a J_2 plasticity model in which the yield stress becomes a function of strain, strain rate, and temperature, but with their relative contributions weighted by constants that have physical meaning. The yield stress is assumed to be

$$\sigma_y = (C_1 + C_2 \varepsilon_p^{\frac{1}{2}}) \exp(-C_3 T + C_4 T \log \dot{\varepsilon}_p) + C_5 \varepsilon_p^n + kl^{-\frac{1}{2}} + \sigma_G. \tag{24}$$

In this relation, $C_1, C_2, C_3, C_4, C_5, k, \sigma_G, l$ are constants. The constants σ_G, k, l represent, respectively, the contributions to the yield stress due to solutes, the initial dislocation density, and the average grain diameter. The remaining constants are selected to distribute the contribution to the hardening of the plastic strain, its rate, and the temperature. Based on the crystallographic structure of the metal under study, some of the constants C_i will be set to zero. For example, fcc metals such as copper will have $C_1 = C_5 = 0$. Iron and other bcc metals will be represented with equations that have $C_2 = 0$. These differences are mainly based on the physical influence of the effects of strain on each type of structure, which is especially dominant when it comes to modeling fcc metals, whereas the strain-rate hardening, thermal softening, and grain size have a greater effect on bcc metals.

2.4.3. Arrhenius-Type Model Constitutive Relations

Last, we consider an Arrhenius-type (AR) constitutive model [19], a strain-compensated equation aiming to reproduce the behavior of metals at high temperature. As in the previous constitutive laws, the AR model is a classical J_2 plasticity model with an elaborated expression for the yield stress σ_y. In this case, it is defined as

$$\sigma_y = \frac{1}{\alpha(\varepsilon_p)} \sinh^{-1} \left(\frac{Z(\varepsilon_p, \dot{\varepsilon}_p, T)}{A(\varepsilon_p)} \right)^{1/n}, \tag{25}$$

where $\alpha : \mathbb{R} \to \mathbb{R}$ and $A : \mathbb{R} \to \mathbb{R}$ are two functions employed to represent the influence of the plastic strain on the response and n is a material exponent. On the other hand, $Z : \mathbb{R} \times \mathbb{R} \times \mathbb{R}^+ \to \mathbb{R}$ is the so-called Zener–Holloman function, accounts for the effects of strain rate $\dot{\varepsilon}_p$ and temperature T, and is defined as

$$Z(\varepsilon_p, \dot{\varepsilon}_p, T) := \dot{\varepsilon}_p \exp \left(\frac{Q(\varepsilon_p)}{RT} \right), \tag{26}$$

where R is the universal gas constant and $Q : \mathbb{R} \to \mathbb{R}$ is the activation energy, assumed to be a third-order polynomial.

The scalar functions that enter the definition of the yield function are thus α, A and Q. The three are defined parametrically as

$$\begin{aligned}
\alpha(\varepsilon_p) &= \alpha_0 + \alpha_1 \varepsilon_p + \alpha_2 \varepsilon_p^2 + \alpha_3 \varepsilon_p^3, \\
Q(\varepsilon_p) &= Q_0 + Q_1 \varepsilon_p + Q_2 \varepsilon_p^2 + Q_3 \varepsilon_p^3, \\
A(\varepsilon_p) &= \exp \left[A_0 + A_1 \varepsilon_p + A_2 \varepsilon_p^2 + A_3 \varepsilon_p^3 \right],
\end{aligned} \tag{27}$$

where $\alpha_0, \ldots, \alpha_3$, $Q_0, \ldots, Q_3$, and $A_0, \ldots, A_3$ are material constants determined experimentally. Depending on the author, these three functions might adopt slightly different forms leading to potentially higher accuracy at the expense of more difficulties for their calibration.

3. Application

The methodology presented in Section 2 is applied now to a relevant example in mechanics of deformable solids, namely, Taylor's impact test [21,22]. In what follows, we will study the calibration of the three material models of Section 2.4 based on the outputs obtained from this well-known test that consists of a high-velocity impact of a metallic anvil onto a rigid wall. As illustrated in Figure 2, the impact creates irrecoverable deformations in the anvil that, due to the symmetry of the problem, can be macroscopically quantified by measuring the changes in the diameter and length of the impactor.

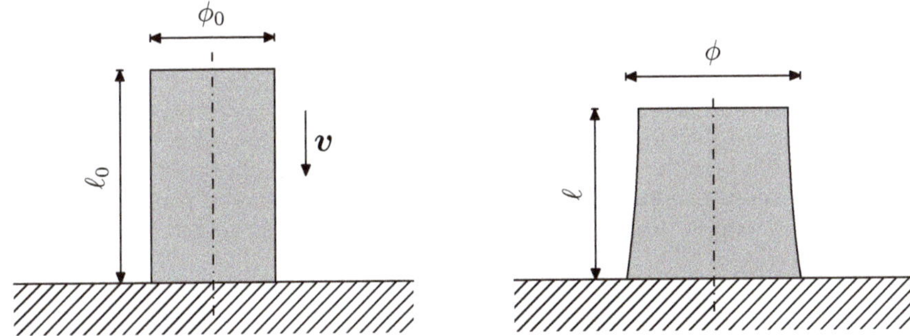

Figure 2. Schematic of Taylor's impact test.

Figure 3 illustrates the procedure advocated for our numerical analysis: starting from a prior distribution for the material parameters, a meta-model of Taylor's impact test is constructed based on anisotropic RBF. The meta-model, once completed, is cheap to run and can be used to perform sensitivity analyses and to update, via Bayesian calibration, the probability distribution of the original parameters. If deemed necessary, the latter probability distribution can be reintroduced in the Bayesian calibration, this time as prior, as illustrated in Figure 3, until the parameter distribution converges to an (almost) stationary function. In theory, one could use the posterior probabilities to start the whole process, helping to build a better meta-model that will be later employed in the GSA and calibration. This route, however, might be too expensive in real life applications.

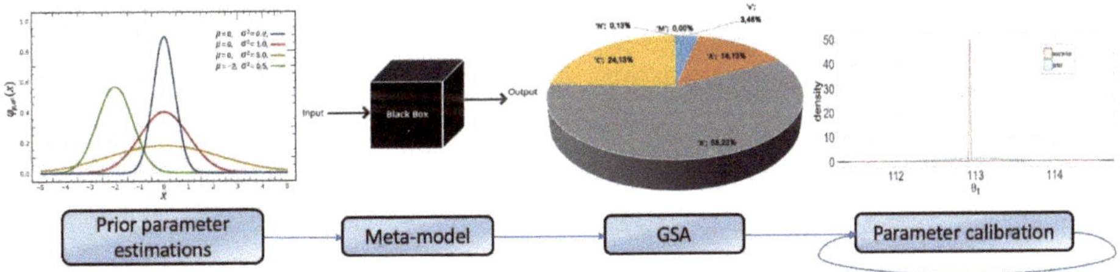

Figure 3. Iterative process for a two-stage approach of screening and calibration of model parameters.

To build the meta-model, five impact velocities are selected over a typical range of Taylor's bar experiments: namely, 200, 230, 260, 290, and 320 m/s. Then, the different tests for each impact velocity are simulated considering a Cr-Mo steel as the anvil's material. Each impact velocity point consists of 612 simulations for the Johnson–Cook and Zerilli–Armstrong models, and 1800 for the Arrherius-type, since the latter involves a larger number of material parameters and requires more data in order to get reliable levels of accuracy when constructing the meta-models. The parameters fed to the simulations have been sampled from uniform distributions centered at nominal values taken from the literature [19,33] with ±10% ranges, varying them according to a Low Discrepancy Design method (LDD), or Sobol sequence [34]. The latter is obtained with a deterministic algorithm that subdivides each dimension of the sample space into 2^N points, while ensuring good uniformity properties.

The QoIs selected for the meta-model are ΔR and ΔL; that is, the changes in radius and length of the anvil after impact. Using the methods described in Section 2.2, an RBF-based meta-model is obtained for each material and impact velocity. The meta-models now serve as the basis for the Global Sensitivity Analysis that will identify the most significant parameters in each model, ruling out from the Bayesian calibration those whose influence on the QoIs is relatively small. Finally, for each of the material models, a full Bayesian analysis will be done based on the concepts of Section 2.3, providing a fitted Gaussian process per model and QoI. This last step demands standard but cumbersome operations and has been performed using a freely available R package [35].

To complete the Bayesian calibration, we need meta-model predictions for arbitrary velocities of the impactor. Since the available meta-models are only defined for five selected velocities, we will interpolate linearly their predictions for the QoI at any intermediate velocity (see Figures 4–6). This strategy will speed-up the generation of data for the Bayesian analysis. To validate it, we will first confirm that the error made by this interpolation is negligible. For that, we will compare solutions obtained with FE simulations at arbitrary velocities of the anvil against interpolated meta-model predictions.

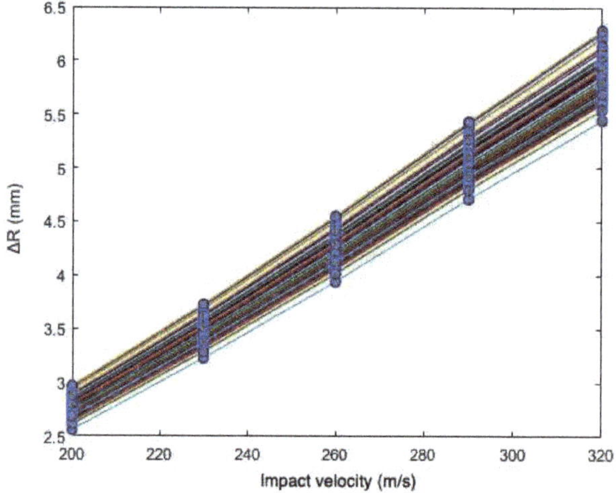

Figure 4. Linear interpolation of meta-model predictions of ΔR for the Johnson–Cook constitutive relation. Each piecewise linear interpolation connects predictions with the same model parameters.

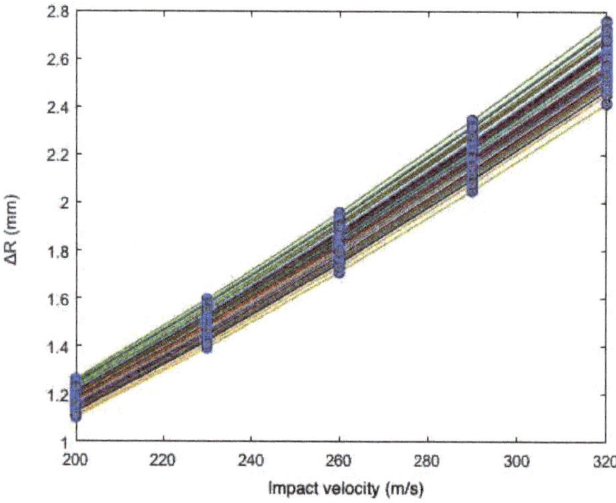

Figure 5. Linear interpolation of meta-model predictions of ΔR for the Zerilli–Armstrong constitutive relation.

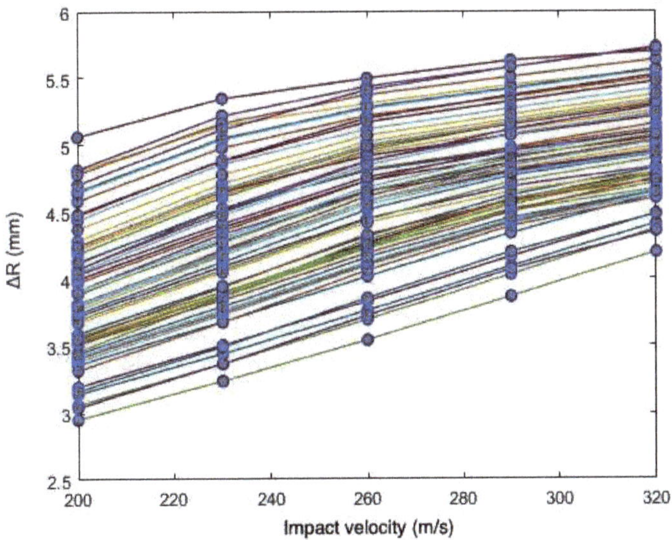

Figure 6. Linear interpolation of meta-model predictions of ΔR for the Arrhenius-type constitutive relation.

Once accepted, this strategy for combining meta-models will result in an extremely cheap source of simulated data that will be used to study the material models. For each of the latter, the fitting data will consist of $n_1 = 20$ sets of observed points $\Delta_1 = \{x_1, \ldots, x_{n_1}\}$, plus $n_2 = 500$ sets of computational outputs, derived from the meta-models interpolation, $\Delta_2 = (x'_1, t_1), \ldots, (x'_{n_2}, t_{n_2})$, where x_i and x'_i are the experimental impact velocities and interpolated impact velocities acting as the variable input, respectively, while t_i are the parameter inputs to the meta-model. To assess the results of the meta-models interpolation, the FE cases against which are to be compared will be generated employing the same parameter inputs t_i and impact velocities x'_i.

In this work, we have chosen to calibrate CrMo steel because the parameters for the JC, ZA, and ARR models could be found in the literature for this material. However, no experimental measurements are available for Taylor tests with anvils of this material. Hence, we follow an alternative avenue to obtain data, one that is often employed in statistical analyses [36,37]. The idea is to generate data from finite element simulations (20 in our procedure) using a fixed material model with nominal parameters, exploring all impact velocities and adding white Gaussian noise to all the measured QoIs, consistent with Equation (18).

To complete the problem definition, it remains to choose prior probability distributions for the complete set of material parameters θ, the variance of the global observation error λ^2, and the hyperparameters χ_δ of the discrepancy function. Tables 1 and 2 describe the probability distributions chosen for each parameter in the three models and the references employed for their choice.

Table 1. Prior probability distributions.

Term/Parameter	Probability Distribution Function
θ	$\mathcal{N}(\mu, 1)$ JC/ARR or $\mathcal{N}(\mu, 10)$ ZA (see Table 2)
λ^2	$\Gamma(1, 0.1)$
σ_δ^2	$\Gamma(1, 0.1)$
ψ_δ	$\mathcal{U}(0, 1)$

Table 2. Mean values for the parameter distributions according to literature [19,33].

Material Model	Mean values
Johnson–Cook	$A = 113$ MPa, $B = 211$ MPa, $C = 0.073$, $n = 0.218$, $m = 0.818$
Zerilli–Armstrong	$C_0 = 707.2$ MPa, $C_1 = 575$ MPa, $C_3 = 0.00698$ K^{-1}, $C_4 = 0.00032$ K^{-1}, $C_5 = 637.5$ MPa, $n = 0.41$
Arrhenius-type	$Q_0 = 412.31$, $Q_1 = -510.82$, $Q_2 = 1873.4$, $Q_3 = -1872.4$ $A_0 = 36.402$, $A_1 = -68.301$, $A_2 = 254.32$, $A_3 = -255.57$ $\alpha_0 = 0.009481$, $\alpha_1 = -0.003841$, $\alpha_2 = -0.012971$, $\alpha_3 = 0.025892$, $n = 5.2248$

4. Results

We now present the results of the GSA analyses, the meta-models interpolation and calibration procedure for the three material models described in Section 2.4 based on the results obtained from the experiments of Taylor's anvil impact. These are obtained from the RBF meta-model whose construction is detailed in Sections 2.2 and 3.

4.1. Sensitivity Analysis

First, we present the results of the sensitivity analyses, as summarized in the pie charts of Figures 7–9. For each of the three material models, these figures depict the contributions, at two impact velocities, of the parameters to global variance, considering independently the two QoIs: namely, the increments in anvil's radius and length.

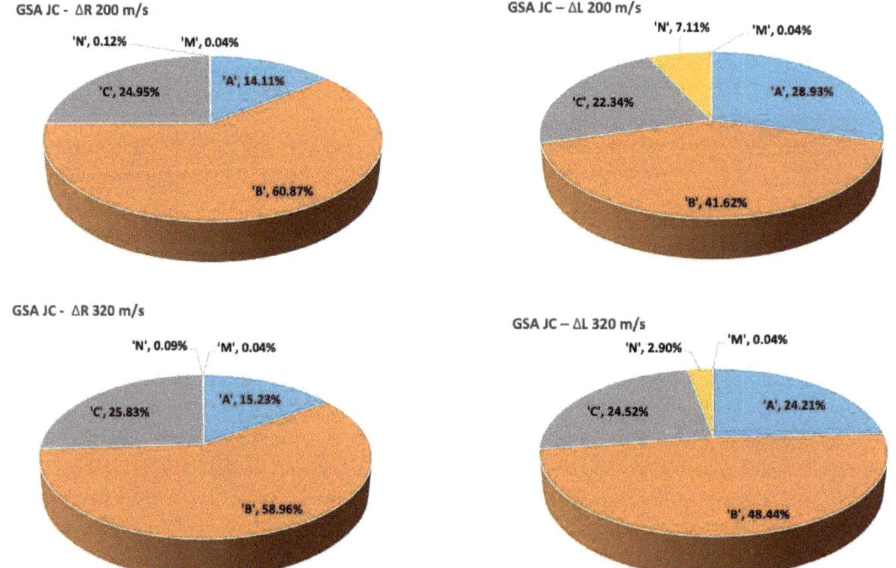

Figure 7. Global Sensitivity Analysis (GSA) results for the Johnson–Cook (JC) model considering ΔR and ΔL at 200 and 320 m/s.

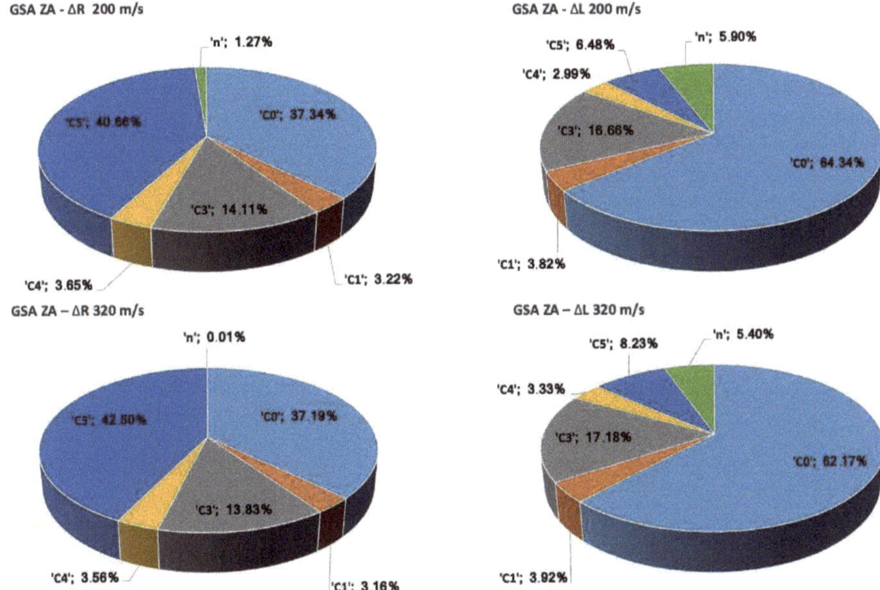

Figure 8. GSA results for the Zerilli–Armstrong (ZA) model considering ΔR and ΔL at 200 and 320 m/s.

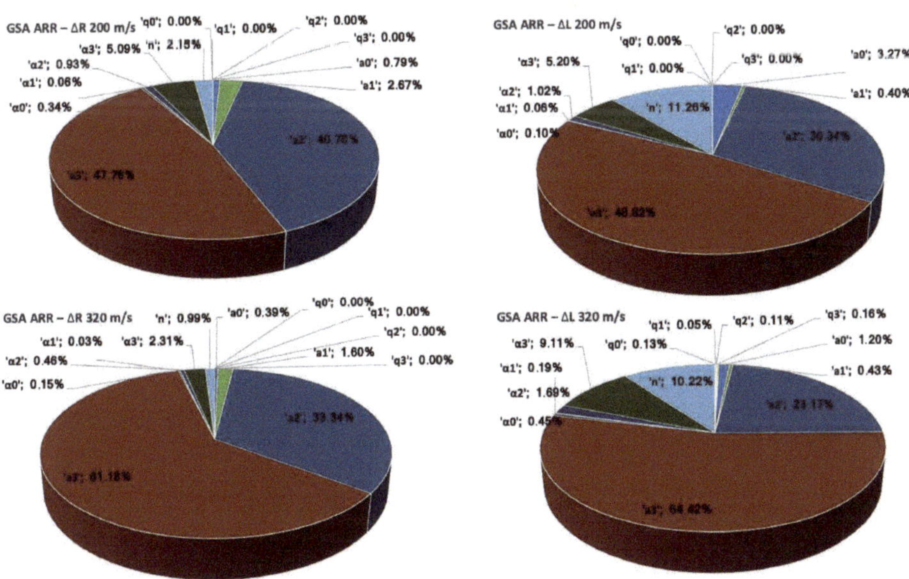

Figure 9. GSA results for the Arrhenius-type model considering ΔR and ΔL at 200 and 320 m/s.

In all the three models, the pie charts expressing the parameters' influence are slightly different, as expected from a complex experiment. However, the most significant result of the analysis performed is that the most influential parameters of each material model coincide in the four sensitivity figures.

To proceed, we identify for each material model the smallest set of parameters whose combined influence accounts for at least 90% of the total QoI variance in all the tests performed and we summarize these findings in Table 3. These results are useful in two ways. First, they simplify the ensuing Bayesian calibration, limiting the number of hyperparameters for the Gaussian processes and the computations involved in the likelihood calculations. Second, from a quantitative point of view, it can be employed by users of material models in numerical simulations to reveal the most influential parameters in the three laws considered, where most of the calibration efforts should be placed, irrespective of the methodology followed to this end.

Table 3. Model parameters accounting for 90% or more of the Quantity of Interest (QoI) variance.

Material Model	Significant Parameters
Johnson–Cook	A, B, C
Zerilli–Armstrong	C_0, C_3, C_5, n
Arrhenius-type	A_2, A_3, α_3, n

4.2. Linear Interpolation of Meta-Models

In Section 3, it was proposed to interpolate linearly meta-model predictions to extend the latter to arbitrary anvil velocities. Next, in Figures 10–12, we show a comparison between the predictions of the QoI ΔR obtained from meta-model interpolation and full FE simulations.

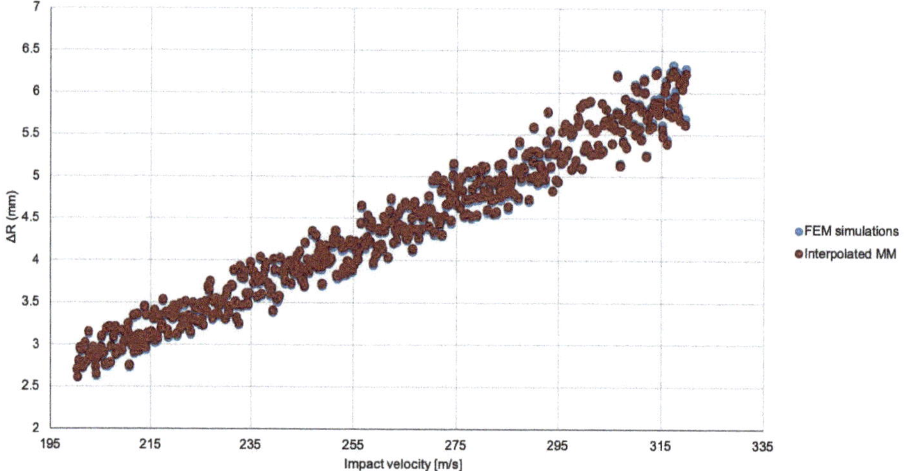

Figure 10. Comparison of meta-models interpolated results and FEM simulations, considering ΔR for the Johnson–Cook model.

Observing these plots and the results collected in Table 4, we conclude that the meta-model interpolation for the Johnson–Cook and Zerilli–Armstrong models provides accurate predictions of ΔR for arbitrary impact velocities. In contrast, the interpolations of the Arrhenius-type model are not as accurate, possibly due to the relatively higher non-linearity of its constitutive equation, affecting directly the flow stress computation. Without a direct means of verifying this assertion, we might speculate that these non-linearities trigger complex deformation patterns in the anvil once the material enters the plastic regime. However, given that the maximum relative error is below 7×10^{-2} in all three cases, we can accept the interpolated predictions for the three constitutive models. This choice will result in huge computational savings for the Bayesian calibration.

Table 4. Errors in the meta-model predictions of ΔR compared with full FE simulations.

Model	Mean Relative Error	Maximum Relative Error
Jonhson–Cook	6.2×10^{-3}	1.2×10^{-2}
Zerilli–Armstrong	8.1×10^{-3}	3.2×10^{-2}
Arrhenius-type	1.1×10^{-2}	6.7×10^{-2}

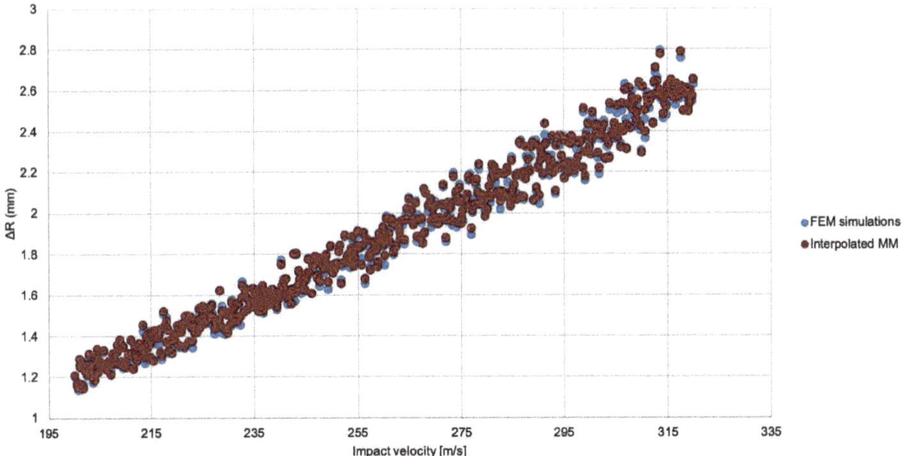

Figure 11. Comparison of meta-models interpolated results and FEM simulation results, considering ΔR for the Zerilli–Armstrong model.

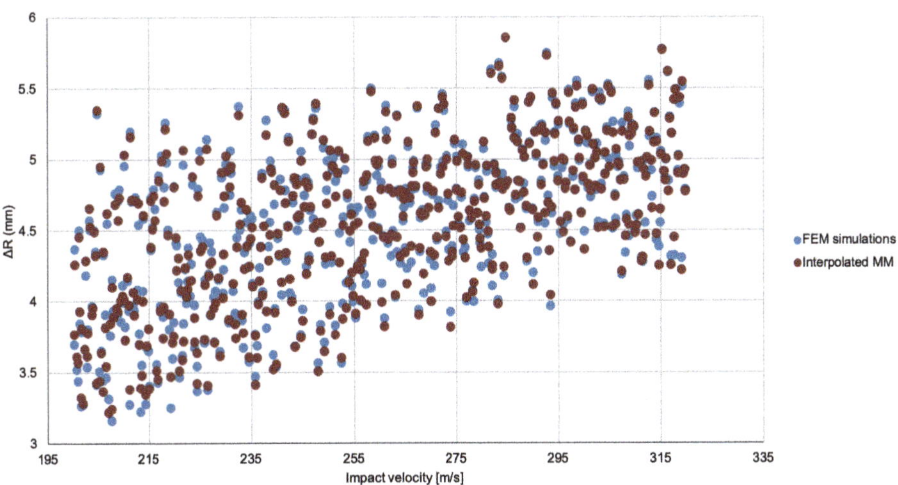

Figure 12. Comparison of meta-models interpolated results and FEM simulation results, considering ΔR for the Arrhenius-type model.

We have also validated the linear interpolation strategy for the quantity ΔL. The results are very similar to the ones obtained for ΔR and the interpolation plots are not presented. Table 5 collects the errors made by the meta-model for ΔL as compared with the FE solution, leading us to conclude, as for the previous QoI, that the interpolated predictions are accurate enough.

Table 5. Errors in the meta-model predictions of ΔL compared with full FE simulations.

Model	Mean Relative Error	Maximum Relative Error
Jonhson–Cook	6.9×10^{-3}	2.3×10^{-2}
Zerilli–Armstrong	9.4×10^{-3}	3.6×10^{-2}
Arrhenius-type	7.8×10^{-3}	3.0×10^{-2}

4.3. Bayesian Calibration

Finally, we proceed to perform a Bayesian calibration of the material models employed in the sensitivity analysis, keeping fixed at their nominal value those parameters that have been found to be non-influential in the sensitivity analyses. The calibration results considering both QoIs, ΔR and ΔL, are shown in Figures 13–18.

Specifically, Figure 13 shows the prior probability distribution functions provided for the three most relevant parameters, $A, B,$ and C of the JC model, and their posterior probability functions. Similarly, Figure 15 illustrates the same probability functions, now for the most relevant parameters of the ZA model: namely, $C_0, C_3, C_5,$ and n.

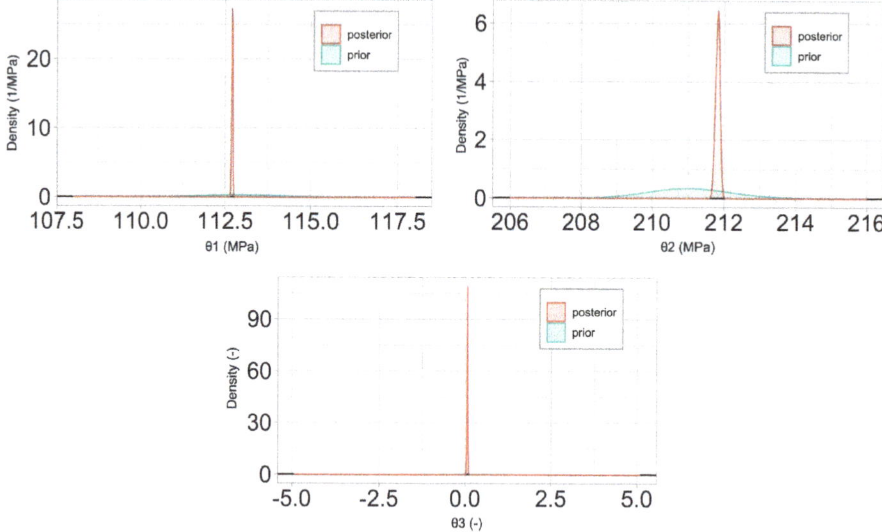

Figure 13. Prior vs. posterior probability distribution functions of parameters $A \sim \theta_1, B \sim \theta_2, C \sim \theta_3,$ in the JC model considering ΔR as the QoI.

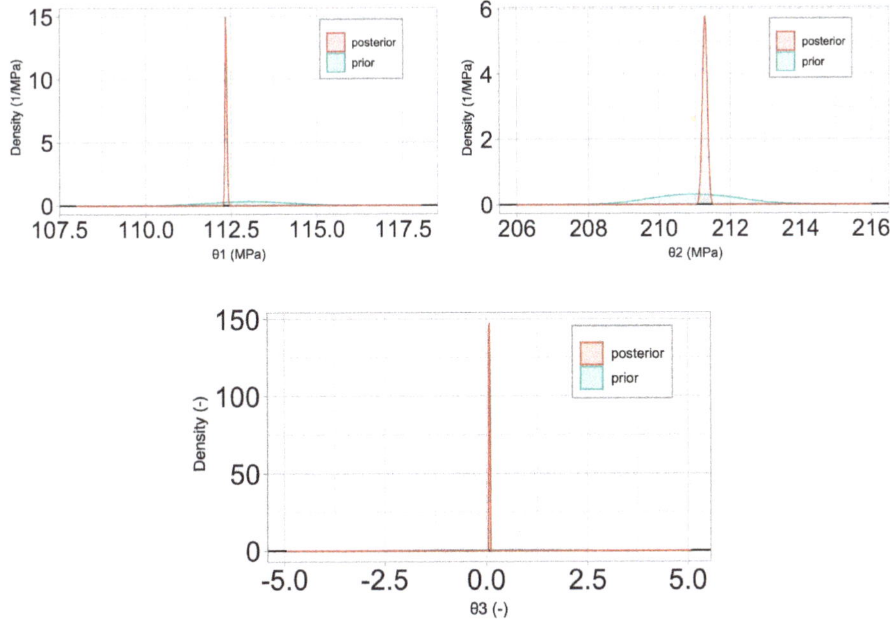

Figure 14. Prior vs. posterior probability distribution functions of parameters $A \sim \theta_1, B \sim \theta_2, C \sim \theta_3$, in the JC model considering ΔL as the QoI.

Based on the results of the calibration, we can make general comments on the calibrated models. In the case of the JC constitutive law, the calibration process has notably sharpened the probability density function of the three most significant parameters (see Figures 13 and 14), eliminating a great part of the uncertainty linked to the variance in the prior probability distributions. Comparing the calibrated values of the parameters in the JC model obtained for the two QoIs analyzed (see Table 6), we note that both are similar. This suggests that the JC model is a good constitutive model for capturing the physics behind Taylor's test. In turn, the posterior probability distributions for the ZA model barely reduce the variance of the priors (cf. Figures 15 and 16). As a consequence, the calibration does not reduces significantly the uncertainty in the parameters. In addition, some of the calibrated parameters for the two QoIs under study have large disparities in their means. This is a consequence of the fact that, in our experiments, simulations carried out with the ZA model predict softer results, irrespective of the impact velocity and QoI observed. This fact, far from being a negative result, proves the potential of the method and illustrates that when the experimental and the simulation data are not in full agreement, the outcome of the calibration alerts of more uncertainties in the model and/or the data, or even the inability of the constitutive model to capture the physics of the problem.

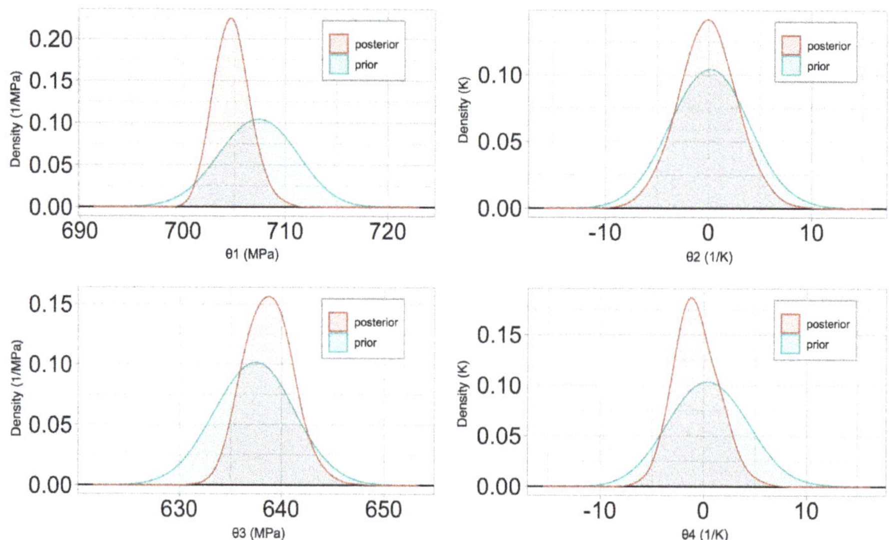

Figure 15. Prior vs. posterior parameter probability distributions for $C_0 \sim \theta_1, C_3 \sim \theta_2, C_5 \sim \theta_3, n \sim \theta_4$, of the ZA model considering ΔR as the QoI.

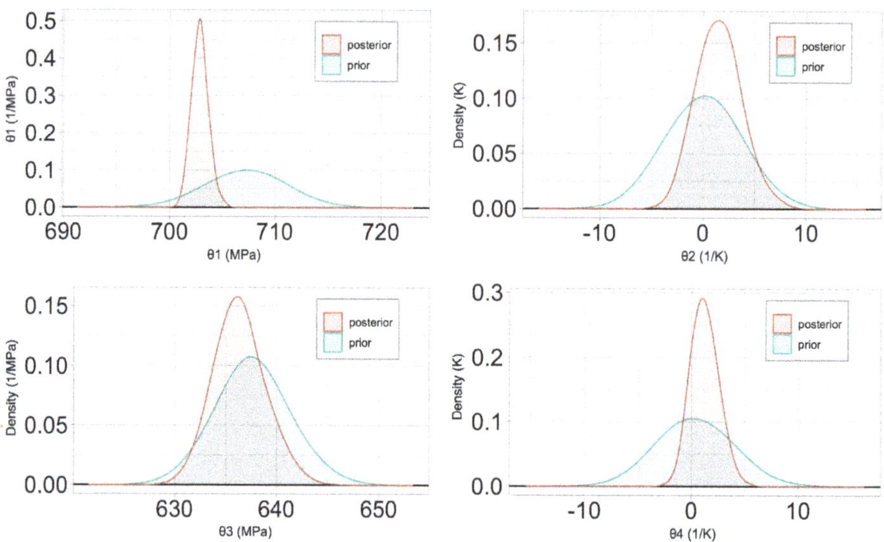

Figure 16. Prior vs. posterior parameter probability distributions for $C_0 \sim \theta_1, C_3 \sim \theta_2, C_5 \sim \theta_3, n \sim \theta_4$, of the ZA model considering ΔL as the QoI.

Regarding the calibration of the Arrhenius model, it can be observed from Figures 17 and 18 that the variance reduction in the posterior probability distribution of its parameters is not as strong as for the Johnson–Cook constitutive law, although it is still significant when compared to the Zerilli–Armstrong case. Something similar happens when analyzing the (mean) calibrated parameters when considering the two QoIs. Even if there is a good agreement among them, the calibrated parameter α_3 obtained for the two QoI is fairly different. A potential explanation can be found, again, in the complexity of the constitutive equation and the effects of ignoring a large number of the model parameters. While this conscious choice saves much computational cost, it causes the loss of information in the model behavior, that, either way, could be countered to a large extent with additional simulated data at a low computational cost.

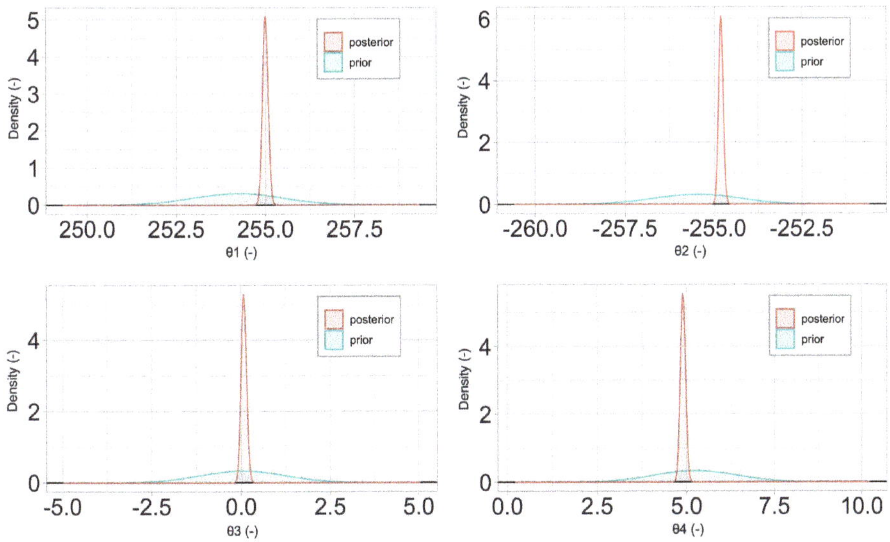

Figure 17. Prior vs. posterior parameter probability distributions for parameters $A_2 \sim \theta_1, A_3 \sim \theta_2, \alpha_3 \sim \theta_3, n \sim \theta_4$ of the Arrhenius-type model considering ΔK as the QoI.

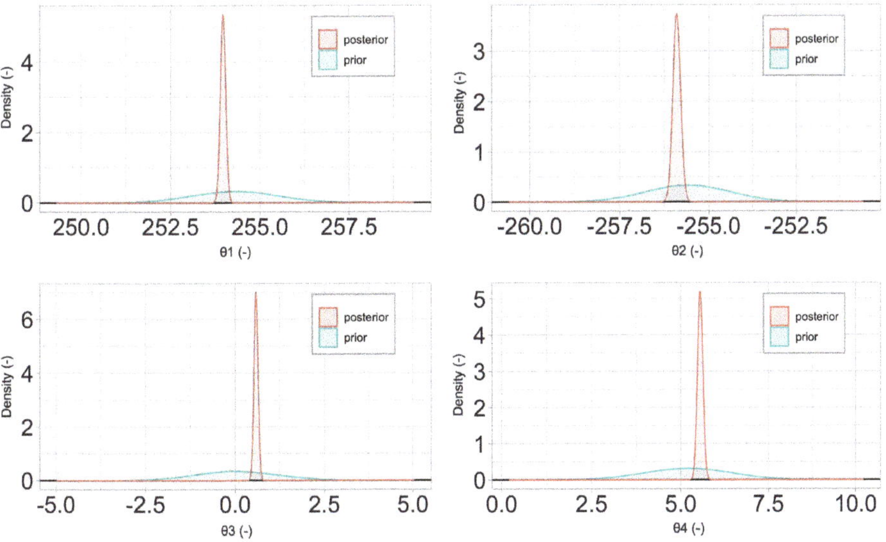

Figure 18. Prior vs. posterior parameter probability distributions for parameters $A_2 \sim \theta_1, A_3 \sim \theta_2, \alpha_3 \sim \theta_3, n \sim \theta_4$ of the Arrhenius-type model considering ΔL as the QoI.

Table 6. A posteriori mean values obtained for each parameter, considering both QoIs, ΔR and ΔL, for the calibration of the models.

Material Model	Mean Values ΔR	Mean Values ΔL
Johnson–Cook	$A = 112.7\,\text{MPa}$, $B = 211.8\,\text{MPa}$, $C = 0.065$	$A = 112.4\,\text{MPa}$, $B = 211.3\,\text{MPa}$, $C = 0.073$
Zerilli–Armstrong	$C_0 = 704.6\,\text{MPa}$, $C_3 = -0.00392\,\text{K}^{-1}$, $C_5 = 638.3\,\text{MPa}$, $n = -1.65$	$C_0 = 702.8\,\text{MPa}$, $C_3 = 1.65\,\text{K}^{-1}$, $C_5 = 636.1\,\text{MPa}$, $n = 1.02$
Arrhenius-type	$A_2 = 255.0$, $A_3 = -254.8$, $\alpha_3 = 0.077$, $n = 4.9$	$A_2 = 254.0$, $A_3 = -255.9$, $\alpha_3 = 0.59$, $n = 5.54$

5. Conclusions

Calibrating complex material models using experimental tests and simulations is a critical task in computational engineering. When done in combination with statistical inference, this process can yield accurate values for the unknown material parameters plus additional information about its scatter and intervals of confidence. For example, Gaussian processes provide a natural and powerful framework to combine physical and numerical tests to obtain probability distributions of the material parameters. To fully exploit the potential of this kind of analysis, however, a large number of data points is required, and the latter can be most effectively obtained by employing a meta-model.

In this work, we described an effective framework for calibrating complex material models based on the combination of meta-models built on top of anisotropic Radial Basis Functions, Global Sensitivity Analysis, and Gaussian processes. The integration of these techniques results in a robust and efficient workflow.

We have employed the framework described for the calibration of three extremely common, although complex material models. These are the Johnson–Cook, the Zerilli–Armstrong, and Arrhenius-type models, and are typically employed for the characterization of the elasto-visco-plastic response of metals under high strain rates, and possibly high temperature as well. The outcome of our analysis is two-fold. First, we are able, for each material model, to rank the sensitivity of an impact simulation with respect to each of the parameters involved. Second, the framework produces a probability distribution for all the calibrated parameters as a function of the available or generated data, tapping into previously built and extremely fast statistical tools to obtain them. Such a characterization is more complete than simple point estimates, often employed when fitting material models.

Let us conclude by noting that the procedures described in this work have applicability beyond materials calibration to, in principle, problems where model evaluations and experimental setups are costly.

Author Contributions: Conceptualization, I.R.; methodology, E.M. and I.R.; software, J.L.d.P., E.M., and I.R.; validation, J.L.d.P. and E.M.; data curation, J.L.d.P. and E.M.; writing—original draft preparation, J.L.d.P. and E.M.; writing—review and editing, I.R.; visualization, J.L.d.P. and E.M.; supervision, I.R.; funding acquisition, I.R. All authors have read and agreed to the published version of the manuscript.

Funding: JdP has been partially funded by the Clean Sky 2 Joint Undertaking under the European Union's Horizon 2020 research and innovation programme (Call Reference No: JTI-CS2-2017-CfP07-ENG-03-22) under grant agreement No 821044. IR would also like to acknowledge the funding received from the Spanish Ministry of Science, Innovation and Universities through project HEXAGB (RTI2018-098245-B- C21).

Acknowledgments: JdP has been partially funded by the Clean Sky 2 Joint Undertaking under the European Union's Horizon 2020 research and innovation programme (Call Reference No: JTI-CS2-2017-CfP07-ENG-03-22) under grant agreement No 821044. IR would also like to acknowledge the funding received from the Spanish Ministry of Science, Innovation and Universities through project HEXAGB (RTI2018-098245-B- C21).

Conflicts of Interest: The authors declare no conflict of interest.

References

1. Iooss, B.; Lemaître, P. A review on global sensitivity analysis methods. In *Uncertainty Management in Simulation-Optimization of Complex Systems*; Springer: Berlin/Heidelberg, Germany, 2015; pp. 101–122.
2. Saltelli, A.; Ratto, M.; Andres, T.; Campolongo, F.; Cariboni, J.; Gatelli, D.; Saisana, M.; Tarantola, S. *Global Sensitivity Analysis: The Primer*; John Wiley & Sons: Hoboken, NJ, USA, 2008.
3. Box, G.; Draper, N. *Empirical Model-Building and Response Surfaces*; John Wiley & Sons: Hoboken, NJ, USA, 1987.
4. Iooss, B.; Van Dorpe, F.; Devictor, N. Response surfaces and sensitivity analyses for an environmental model of dose calculations. *Reliab. Eng. Syst. Saf.* **2006**, *91*, 1241–1251.
5. Rohmer, J. Dynamic sensitivity analysis of long-running landslide models through basis set expansion and meta-modelling. *Nat. Hazards* **2014**, *73*, 5–22.
6. Todri, E.; Amenaghawon, A.; Del Val, I.; Leak, D.; Kontoravdi, C.; Kucherenko, S.; Shah, N. Global sensitivity analysis and meta-modeling of an ethanol production process. *Chem. Eng. Sci.* **2014**, *114*, 114–127.
7. Welch, W.; Buck, R.; Sacks, J.; Wynn, H.; Mitchell, T.; Morris, M. Screening, predicting, and computer experiments. *Technometrics* **1992**, *34*, 15–25.
8. Buhmann, M. *Radial Basis Functions: Theory and Implementations*; Cambridge University Press: Cambridge, UK, 2003; Volume 12.
9. MacAllister, A.; Kohl, A.; Winer, E. Using High-Fidelity Meta-Models to Improve Performance of Small Dataset Trained Bayesian Networks. *Expert Syst. Appl.* **2019**, *139*, 112830.
10. Guttman, I.; Tiao, G. *A Bayesian Approach to Some Best Population Problems*; Technical Report; Wisconsin Univ-Madison: Madison, WI, USA, 1963.
11. Aitchison, J.; Dunsmore, I. *Statistical Prediction Analysis*; CUP Archive: Cambridge, UK, 1980.
12. Kennedy, M.; O'Hagan, A. Bayesian calibration of computer models. *J. R. Stat. Soc. Ser. B Stat. Methodol.* **2001**, *63*, 425–464.
13. Higdon, D.; Kennedy, M.; Cavendish, J.; Cafeo, J.; Ryne, R. Combining Field Data and Computer Simulations for Calibration and Prediction. *SIAM J. Sci. Comput.* **2004**, *26*, 448–466.
14. OHagan, A. Bayesian analysis of computer code outputs: A tutorial. *Reliab. Eng. Syst. Saf.* **2006**, *91*, 1290–1300.
15. Draper, D.; Pereira, A.; Prado, P.; Saltelli, A.; Cheal, R.; Eguilior, S.; Mendes, B.; Tarantola, S. Scenario and parametric uncertainty in GESAMAC: A methodological study in nuclear waste disposal risk assessment. *Comput. Phys. Commun.* **1999**, *117*, 142–155.
16. Janse, J.; Scheffer, M.; Lijklema, L.; Van Liere, L.; Sloot, J.; Mooij, W. Estimating the critical phosphorus loading of shallow lakes with the ecosystem model PCLake: Sensitivity, calibration and uncertainty. *Ecol. Model.* **2010**, *221*, 654–665.
17. Johnson, G. A constitutive model and data for materials subjected to large strains, high strain rates, and high temperatures. In Proceedings of the 7th International Symposium on Ballistics, Hague, The Netherlands, 19–21 April 1983; pp. 541–547.
18. Zerilli, F.; Armstrong, R. Dislocation-mechanics-based constitutive relations for material dynamics calculations. *J. Appl. Phys.* **1987**, *61*, 1816–1825.
19. Samantaray, D.; Mandal, S.; Bhaduri, A. A comparative study on Johnson–Cook, modified Zerilli–Armstrong and Arrhenius-type constitutive models to predict elevated temperature flow behaviour in modified 9Cr–1Mo steel. *Comput. Mater. Sci.* **2009**, *47*, 568–576.
20. Portillo, D.; del Pozo, D.; Rodríguez-Galán, D.; Segurado, J.; Romero, I. MUESLI—A Material UnivErSal LIbrary. *Adv. Eng. Softw.* **2017**, *105*, 1–8.
21. Taylor, G. The use of flat-ended projectiles for determining dynamic yield stress I. Theoretical considerations. *Proc. R. Soc. Lond. Ser. A Math. Phys. Sci.* **1948**, *194*, 289–299.
22. Whiffin, A. The use of flat-ended projectiles for determining dynamic yield stress-II. Tests on various metallic materials. *Proc. R. Soc. Lond. Ser. A Math. Phys. Sci.* **1948**, *194*, 300–322.
23. Menga, E.; Sánchez, M.; Romero, I. Anisotropic meta-models for computationally expensive simulations in nonlinear mechanics. *Int. J. Numer. Methods Eng.* **2019**, doi:10.1002/nme.6250.
24. Hoff, P. *A First Course in Bayesian Statistical Methods*; Springer: Berlin/Heidelberg, Germany, 2009; Volume 580.
25. Wernick, M.; Yang, Y.; Brankov, J.; Yourganov, G.; Strother, S. Machine learning in medical imaging. *IEEE Signal Process. Mag.* **2010**, *27*, 25–38.

26. Koenig, N.; Matarić, M. Robot life-long task learning from human demonstrations: A Bayesian approach. *Auton. Robot.* **2017**, *41*, 1173–1188.

27. Cofino, A.; Cano Trueba, R.; Sordo, C.; Gutiérrez Llorente, J. Bayesian networks for probabilistic weather prediction. In Proceedings of the 15th Eureopean Conference on Artificial Intelligence, ECAI'2002, Lyon, France, 21–26 July 2002.

28. Rasmussen, C.; Williams, C.K. *Gaussian Processes for Machine Learning*; MIT Press: Cambridge, MA, USA, 2006.

29. Carmassi, M.; Barbillon, P.; Keller, M.; Parent, E.; Chiodetti, M. Bayesian calibration of a numerical code for prediction. *arXiv* **2018**, arXiv:1801.01810.

30. Shrot, A.; Bäker, M. Determination of Johnson–Cook parameters from machining simulations. *Comput. Mater. Sci.* **2012**, *52*, 298–304.

31. Li, H.; He, L.; Zhao, G.; Zhang, L. Constitutive relationships of hot stamping boron steel B1500HS based on the modified Arrhenius and Johnson–Cook model. *Mater. Sci. Eng. A* **2013**, *580*, 330–348.

32. Banerjee, A.; Dhar, S.; Acharyya, S.; Datta, D.; Nayak, N. Determination of Johnson Cook material and failure model constants and numerical modelling of Charpy impact test of armour steel. *Mater. Sci. Eng. A* **2015**, *640*, 200–209.

33. Valentin, T.; Magain, P.; Quik, M.; Labibes, K.; Albertini, C. Validation of constitutive equations for steel. *J. Phys. IV* **1997**, *7*, C3–611.

34. Sobol', I. On the distribution of points in a cube and the approximate evaluation of integrals. *Zhurnal Vychislitel'noi Matematiki i Matematicheskoi Fiziki* **1967**, *7*, 784–802.

35. Carmassi, M.; Barbillon, P.; Chiodetti, M.; Keller, M.; Parent, E. CaliCo: An R package for Bayesian calibration. *arXiv* **2018**, arXiv:1808.01932.

36. Craig, P.; Goldstein, M.; Rougier, J.; Seheult, A. Bayesian forecasting for complex systems using computer simulators. *J. Am. Stat. Assoc.* **2001**, *96*, 717–729.

37. Riddle, M.; Muehleisen, R. A guide to Bayesian calibration of building energy models. In Proceedings of the Building Simulation Conference, Atlanta, GA, USA, 10–12 September 2014.

 *materials*

Article

Data-Oriented Constitutive Modeling of Plasticity in Metals

Alexander Hartmaier

ICAMS, Ruhr-Universität Bochum, 44801 Bochum, Germany; alexander.hartmaier@rub.de

Received: 9 March 2020; Accepted: 27 March 2020; Published: 1 April 2020

Abstract: Constitutive models for plastic deformation of metals are typically based on flow rules determining the transition from elastic to plastic response of a material as function of the applied mechanical load. These flow rules are commonly formulated as a yield function, based on the equivalent stress and the yield strength of the material, and its derivatives. In this work, a novel mathematical formulation is developed that allows the efficient use of machine learning algorithms describing the elastic-plastic deformation of a solid under arbitrary mechanical loads and that can replace the standard yield functions with more flexible algorithms. By exploiting basic physical principles of elastic-plastic deformation, the dimensionality of the problem is reduced without loss of generality. The data-oriented approach inherently offers a great flexibility to handle different kinds of material anisotropy without the need for explicitly calculating a large number of model parameters. The applicability of this formulation in finite element analysis is demonstrated, and the results are compared to formulations based on Hill-like anisotropic plasticity as reference model. In future applications, the machine learning algorithm can be trained by hybrid experimental and numerical data, as for example obtained from fundamental micromechanical simulations based on crystal plasticity models. In this way, data-oriented constitutive modeling will also provide a new way to homogenize numerical results in a scale-bridging approach.

Keywords: plasticity; machine learning; constitutive modeling

1. Introduction

Finite element analysis (FEA) is a widespread numerical tool for studying the mechanical behavior of structures. While in many applications it is sufficient to know under which conditions a part of the structure fails plastically or suffers damage or fracture, in some cases, like in sheet forming or for crash simulations, it is important to be able to simulate the plastic deformation during loading and to obtain the shape of the structure after the external load is released. Such non-linear behavior is typically described by constitutive models that relate stress and strain in a material, as described in any textbook on non-linear finite element modeling, e.g., see [1]. Conventionally, constitutive relations for plasticity are formulated as flow rules based on a plastic potential. In the simplest case, the latter is the yield function of the material, determining at which local stress the material starts yielding plastically and which plastic strain increment will result from such plastic deformation. As described in the next section, such yield functions relate the equivalent stress and the yield strength of a material, which needs to be determined experimentally or with the help of more fundamental models in a scale-bridging approach.

Experimentally, the yield strength is typically determined in uniaxial tensile tests, in which materials frequently exhibit an anisotropy in their plastic behavior, i.e., the yield strength depends on the orientation of the loading axis with respect to the material coordinate system, defined for example by rolling, normal, and transverse direction. In conventional approaches such anisotropic plastic behavior is described by defining a proper equivalent stress that takes into account material anisotropy, such that yielding occurs at a constant scalar yield strength. This approach has been introduced by Hill [2] and applied to orthotropic plasticity in sheet metals [3]. The concept has been generalized to linear transformation-based anisotropic yield functions by Barlat et al. [4] and to methods describing distortions of the yield surface caused by anisotropic work hardening [5]. In forming technology, similar ideas have been successfully applied to predict the resulting shape of sheet metals after deep drawing [6]. All these approaches have in common that the information about the material anisotropy is mapped into the definition of the equivalent stress, while they differ in the amount of material parameters that is required to describe the anisotropy in the material's flow behavior. To determine these parameters, a series of experiments with different mutual orientations of loading axis and material axis is necessary. Alternatively, micromechanical models, in which discrete representations of the material's microstructure are used together with quite fundamental crystal plasticity models, can be used to calculate the anisotropic flow behavior of a polycrystalline metal [7,8].

To model non-linear material behavior, in more recent approaches, the method of data-based mechanics has been introduced by Kirchdoerfer and Ortiz [9], in which stress-strain data from experimental tests are used directly, rather than using constitutive rules. While the first approaches have been limited to elastic structures under static loads, recently, this concept has been extended to dynamics [10] and to inelastic material behavior [11]. These methods are based on a fundamental re-formulation of the basic equations of mechanics and thus require completely new mechanical solvers. Other data-driven methods in plasticity are formulated as process models, e.g., for air-bending [12], or focus on the application of data-oriented methods as constitutive models in computational plasticity [13]. The latter idea allows the use of existing FEA solvers for mechanical problems, and is also followed in this work, in which a new formulation of a data-oriented flow rule is introduced that can replace conventional constitutive models—formulated in a mathematical closed form—by machine learning (ML) algorithms. In this data-oriented formulation, the anisotropy of the material's flow behavior is considered in a directionally dependent yield strength of the material rather than by an anisotropic transformation of the stress. Using ML algorithms as yield functions provides a great flexibility to describe arbitrary mathematical functions, and at the same time, holds the potential to handle large data sets and multi-dimensional feature vectors as input. Hence, using ML algorithms as constitutive rules for plastic material behavior offers the possibility to explicitly take into account the microstructural information of the material in constitutive modeling. Furthermore, data resulting from experiment and micromechanical simulations can be hybridized to generate training data sets. An overview on applications of ML and data-mining methods in continuum mechanical simulations of material behavior has been provided by Bock et al. [14].

ML algorithms can be classified into algorithms for supervised and unsupervised learning. The former group can be further categorized into classification algorithms, which divide a multi-dimensional feature space into regions with similar properties, and into regressors, which provide linear or non-linear regression functions for the given multi-dimensional data set. Support Vector Machines (SVM) are successfully applied both as classifiers (SVC) [15] and regressors (SVR) [16,17]; an overview on both applications is given in a technical report by Gunn [18]. Since yield functions in continuum plasticity are also employed to subdivide stress space into elastic and plastic regions, this work aims at investigating the possibility of using SVC for the purpose of constitutive modeling in plasticity. An overview on data-mining methods and statistical learning, also covering the SVM method, is given by Hastie et al. [19].

The present paper is organized as follows: In the next section, the basic concepts of continuum plasticity are briefly summarized, and a new mathematical formulation is introduced, which enables a data-oriented approach to constitutive modeling. Subsequently, a consistent formulation of the SVC method to serve as yield function in continuum plasticity is introduced, which is then trained with artificial data resulting from a Hill-like reference material. Using such data has the advantage of being able to judge the quality of the approximations in an objective way. In the next step, trained ML yield function is applied as constitutive model in simple finite element simulations (Supplementary

Materials) to demonstrate its applicability for this purpose. Finally, the results of ML and conventional flow rules are compared and the conclusions drawn from this comparison are presented.

2. Methods

2.1. Anisotropic Continuum Plasticity

In order to describe the elastic-plastic deformation of a material, we introduce the strain tensor ϵ that describes the deformation of the material and the stress tensor σ that describes the forces acting on the surface of the material. Note that tensorial quantities with rank≥ 1 are typeset in bold letters, whereas scalar quantities are represented by standard characters. In the elastic regime, Hooke's law is used as constitutive relation between stress and strain, such that

$$\sigma = C\epsilon, \tag{1}$$

where C is the fourth-rank elasticity tensor of the material. To describe plastic deformation, the yield function of the material is introduced as

$$f(\sigma) = \sigma_{\text{eq}} - \sigma_y, \tag{2}$$

which takes negative values if the equivalent stress σ_{eq} is smaller than the yield strength σ_y of the material, i.e., in the elastic regime. When $f = 0$ plastic yielding sets in, and in case of work hardening, σ_y should be considered as flow stress after this point. Since this work only deals with the onset of plastic yielding, ideal plasticity will be assumed throughout, such that σ_y is a constant, irrespective of the deformation history of the material. Denoting the principal stresses of the stress tensor σ as σ_j with $(j = 1, 2, 3)$, the equivalent stress takes the form

$$\sigma_{eq}^{J2} = \sqrt{\frac{1}{2}\left[(\sigma_1 - \sigma_2)^2 + (\sigma_2 - \sigma_3)^2 + (\sigma_3 - \sigma_1)^2\right]}, \tag{3}$$

which—following the definition of von Mises (see, e.g., the translation of the original work by D. H. Delphenich [20])—is based on the second invariant of the stress deviator (J2). In conjunction with the yield function of Equation (2), it describes the onset of plastic yielding for isotropic materials. Note that the formulation in Equation (3) is intrinsically independent of hydrostatic stress components $p = 1/3\text{Trace}(\sigma)$ and thus does not require to explicitly calculate the deviatoric stress

$$\sigma' = \sigma - pU, \tag{4}$$

where U is the unit tensor. By this definition of the equivalent stress, it is inherently assumed that hydrostatic stress components do not affect the plastic flow behavior of the considered material, which is typically fulfilled for metals, but not for polymers or rocks, such that the method formulated here, will mainly apply to metallic materials or, more generally, to materials, where hydrostatic stresses do not influence the plastic behavior.

As described in the introduction, many materials exhibit a directionally dependent yield strength, such that anisotropic flow rules need to be introduced. A first definition of such anisotropic flow rules was introduced by Hill [2], who used a generalized definition of the equivalent stress to achieve a directionally dependent mapping of the equivalent stresses to maintain a constant yield strength. Hence, in this formulation, the anisotropy is considered in the stress rather than in the yield strength. Since the mathematical formulations in this work are purely based on principal stresses, we use a simplified version of the Hill definition and introduce a Hill-like anisotropic definition of the equivalent stress as

$$\sigma_{\text{eq}} = \sqrt{\frac{1}{2}\left[H_1\,(\sigma_1 - \sigma_2)^2 + H_2\,(\sigma_2 - \sigma_3)^2 + H_3\,(\sigma_3 - \sigma_1)^2\right]}, \tag{5}$$

with only three material parameters H_1, H_2 and H_3, whereas in his original work, Hill introduced three more parameters for an orthotropic material to scale also the shear stress components. Since for orthotropic materials loaded along the main material axes there is no mutual influence of shear and normal components of stress and strain, the formulation introduced here is restricted to loading situations that only produce normal stresses and strains, and where consequently all off-diagonal components of stress and strains tensors remain zero. Furthermore, it is assumed that the loading axes and the main axes of the orthotropic material coincide. Hence, this formalism is currently only valid for a small subset of loading conditions for materials with orthotropic flow anisotropy. The definition of the equivalent stress following Hill can be considered as a generalization of the J2 equivalent stress, because for isotropy, i.e., $H_1 = H_2 = H_3 = 1$, both definitions are equal.

The restrictions applied in this work, allow the mathematical notation to be simplified by only considering principal stresses. In future work, it is intended to render the formulation more general by exploiting that for any stress state, there exists a coordinate system in which the given stress tensor becomes a diagonal tensor composed of the principal stresses σ_j. This coordinate system is given by the eigenvectors of the stress, representing the principal directions, such that the coordinate system of the original stress tensor—and with it the material axes—can be rotated into the coordinate system of the eigenvectors of the stress tensor. In this orientation the stress tensor becomes a diagonal tensor, and Equation (5) can be evaluated with parameters H'_i in the rotated state of the material axes.

The thus defined yield function can be used to determine whether a given stress state results in a purely elastic or rather in an elastic-plastic deformation of a material. The condition $f(\boldsymbol{\sigma}) = 0$ relates stresses lying on a specific hyperplane in stress space, the so-called the yield-locus. Since a material does not sustain any stresses larger than the yield stress (for ideal plasticity) or the flow stress (in case of work hardening), acceptable stress states either produce a negative value of the yield function (elasticity) or lie on the yield locus (plasticity), which should be a convex hull of the elastic stress states. Hence, if a predictor step in finite element analysis (FEA) produces a stress outside the yield locus, a plastic strain increment must be calculated that leads again to an accepted stress state on the yield locus. The return mapping algorithm to calculate such strain increments has been described in many text books on continuum plasticity and non-linear FEA, such that here only a very brief summary based on [1] is reproduced. According to the Prandtl–Reuss flow rule, the plastic strain increment for a given time step can be calculated as

$$\dot{\boldsymbol{\epsilon}}_p = \lambda \frac{\partial f}{\partial \boldsymbol{\sigma}} = \lambda \boldsymbol{n}, \tag{6}$$

where $\boldsymbol{n}$ is the normal vector to the yield locus, defined by the gradient of the yield function $\partial f / \partial \boldsymbol{\sigma}$, and $\lambda > 0$ is the so-called plastic strain multiplier that can be evaluated as

$$\lambda = \frac{\boldsymbol{n} \cdot \boldsymbol{C} \dot{\boldsymbol{\epsilon}}}{\boldsymbol{n} \cdot \boldsymbol{C} \boldsymbol{n}}, \tag{7}$$

where $\dot{\boldsymbol{\epsilon}}$ is the total strain increment of the FEA predictor step that leads to a stress state outside the yield locus and which is consequently decomposed into the plastic strain increment, given by Equation (6), and the elastic strain increment or stress increment given by

$$\dot{\boldsymbol{\sigma}} = \boldsymbol{C}_t \dot{\boldsymbol{\epsilon}} \tag{8}$$

with the tangent stiffness tensor

$$\boldsymbol{C}_t = \boldsymbol{C} - \frac{\boldsymbol{C} \boldsymbol{n} \otimes \boldsymbol{C} \boldsymbol{n}}{\boldsymbol{n} \cdot \boldsymbol{C} \boldsymbol{n}} \tag{9}$$

where "$\otimes$" denotes the tensorial product in the form $a_i \otimes b_j = a_i b_j$.

The gradient of the yield function with respect to the principal stresses can be evaluated analytically as

$$\frac{\partial f}{\partial \sigma_1} = \frac{\partial \sigma_{eq}}{\partial \sigma_1} = \frac{(H_1 + H_3)\,\sigma_1 - H_1\sigma_2 - H_3\sigma_3}{\sigma_{eq}}$$
$$\frac{\partial f}{\partial \sigma_2} = \frac{\partial \sigma_{eq}}{\partial \sigma_2} = \frac{(H_2 + H_1)\,\sigma_2 - H_1\sigma_1 - H_2\sigma_3}{\sigma_{eq}} \tag{10}$$
$$\frac{\partial f}{\partial \sigma_3} = \frac{\partial \sigma_{eq}}{\partial \sigma_3} = \frac{(H_3 + H_2)\,\sigma_3 - H_3\sigma_1 - H_2\sigma_2}{\sigma_{eq}}$$

Note that in the case of isotropic plasticity ($H_1 = H_2 = H_3 = 1$), the gradient takes the simple form

$$\frac{\partial f}{\partial \boldsymbol{\sigma}} = 3\frac{\boldsymbol{\sigma} - p\mathbf{U}}{\sigma_{eq}} = 3\frac{\boldsymbol{\sigma}'}{\sigma_{eq}}. \tag{11}$$

This section served the purpose to introduce the main physical quantities in the notation used in this work. For further details of continuum plasticity or FEA, the reader is referred to standard textbooks, as for example [1]. In the following, the formalism for the data-oriented constitutive model based on a machine learning (ML) yield function is laid out.

2.2. Stress Space in Cylindrical Coordinates

Since plastic deformation in most metals does not depend on hydrostatic stress components, it is useful to transform principal stresses from their representation as a 3-dimensional (3D) Cartesian vector of principal stresses $\boldsymbol{\sigma} = (\sigma_1, \sigma_2, \sigma_3)$ into a cylindrical coordinate system with $\boldsymbol{s} = (\sigma_{eq}, \theta, p)$, where the equivalent stress σ_{eq} represents the norm of the stress deviator $\boldsymbol{\sigma}'$, and the polar angle θ lies in the deviatoric plane normal to the hydrostatic axis p, which has already been used by Hill [2]. This coordinate transformation improves the efficiency of the training, because only two-dimensional data for the equivalent stress and the polar angle need to be used as training features, whereas the hydrostatic component is disregarded. Hence, by exploiting basic physical principles, we effectively reduce the dimensionality of the problem from 6 independent components of an arbitrary stress tensor to 2 degrees of freedom, without loosing the generality of the formulation. As the polar angle θ can be considered a generalized Lode angle [21], it is noted that the Lode angle, by definition, describes the axiality of a loading state in a way that uniaxial loads in different directions result in the same Lode angle. Since our formulation aims at describing anisotropy in the plastic deformation, uniaxial stresses in different directions must possess different angles. To achieve this, we introduce a complex-valued deviatoric stress

$$\sigma_c' = \boldsymbol{\sigma} \cdot \boldsymbol{a} + i\boldsymbol{\sigma} \cdot \boldsymbol{b} = \sqrt{2/3}\,\sigma_{eq}\,e^{i\theta}, \tag{12}$$

where i is the imaginary unit, such that the polar angle

$$\theta = \arg \sigma_c' = -i\,\ln \frac{\boldsymbol{\sigma} \cdot \boldsymbol{a} + i\boldsymbol{\sigma} \cdot \boldsymbol{b}}{\sqrt{2/3}\,\sigma_{eq}}. \tag{13}$$

with the unit vectors $\boldsymbol{a} = (2, -1, -1)/\sqrt{6}$ and $\boldsymbol{b} = (0, 1, -1)/\sqrt{2}$ that span the plane normal to the hydrostatic axis $\boldsymbol{c} = (1, 1, 1)/\sqrt{3}$.

To transform the gradient of the yield function from this cylindrical stress space back to the principle stress space, in which form it is used to calculate the direction of the plastic strain increments in the return mapping algorithm of the plasticity model, we introduce the Jacobian matrix for this coordinate transformation as

$$J = \frac{\partial \boldsymbol{s}}{\partial \boldsymbol{\sigma}} = \begin{pmatrix} \frac{\partial \sigma_{eq}}{\partial \sigma_1} & \frac{\partial \theta}{\partial \sigma_1} & \frac{\partial p}{\partial \sigma_1} \\ \frac{\partial \sigma_{eq}}{\partial \sigma_2} & \frac{\partial \theta}{\partial \sigma_2} & \frac{\partial p}{\partial \sigma_1} \\ \frac{\partial \sigma_{eq}}{\partial \sigma_3} & \frac{\partial \theta}{\partial \sigma_3} & \frac{\partial p}{\partial \sigma_1} \end{pmatrix} \tag{14}$$

where $\partial\sigma_{\mathrm{eq}}/\partial\sigma_j$ is given in Equation (10), $\partial p/\partial\sigma_j = 1/3$ and

$$\frac{\partial\theta}{\partial\sigma} = -i\left(\frac{a+ib}{\sigma\cdot a + i\sigma\cdot b} - \frac{3\sigma'}{\sigma_{\mathrm{eq}}^2}\right). \tag{15}$$

With this Jacobian, the gradient can be calculated as

$$\frac{\partial f}{\partial\sigma} = J\frac{\partial f}{\partial s}. \tag{16}$$

2.3. Data-Oriented Yield Function

While the concept of mapping the equivalent stress in describing anisotropic flow behavior has been applied successfully in the approaches of Hill [2,3] and Barlat [4], for a data-oriented yield function, it is impracticable to calculate the necessary parameters for this stress mapping explicitly. Hence, it is of advantage to reformulate the flow rule in such a way that the yield strength is considered to be directionally dependent, whereas the equivalent stress is formulated in an objective way, without prior knowledge of the material behavior. This is achieved by using the J2 equivalent stress in the flow rule and formally considering the flow stress to be a function of the polar angle in the deviatoric plane, such that

$$f_d(s) = s_1 - \sigma_y(s_2) = \sigma_{\mathrm{eq}}^{\mathrm{J2}} - \sigma_y(\theta). \tag{17}$$

A further advantage of this formulation is that the dependence of the yield function on the two degrees of freedom of the cylindrical stress notation is separated into two independent terms. Furthermore, for symmetry reasons, it is required that the yield strength is a periodic function of the polar angle with periodicity 2π. The gradient of the ML yield function w.r.t. the cylindrical coordinates reads

$$\frac{\partial f_d}{\partial s_1} = \frac{\partial f_d}{\partial\sigma_{\mathrm{eq}}^{\mathrm{J2}}} = 1$$

$$\frac{\partial f_d}{\partial s_2} = \frac{\partial f_d}{\partial\theta} = \frac{d\sigma_y}{d\theta}$$

$$\frac{\partial f_d}{\partial s_3} = \frac{\partial f_d}{\partial p} = 0.$$

It is seen that in the cylindrical stress space $\partial f_d/\partial\sigma_{\mathrm{eq}} = 1$ and $\partial f_d/\partial p = 0$, under the condition that plasticity is independent of hydrostatic stress components. Hence, the only non-constant component of the gradient is $\partial f_d/\partial\theta$, which simplifies the numerical implementation of the method. For isotropic J2 plasticity, $\partial f_d/\partial\theta = 0$, and in this case it is particularly easy to calculate the gradient and to see that the formulations in both coordinate systems result in the same gradient. The transformation of this gradient into the principal stress space is achieved by multiplication with the Jacobian, according to Equation (16).

To establish a data-oriented formulation, we introduce a yield function in the form of a machine learning (ML) algorithm, rather than in a mathematically closed form with a number of model parameters that need to be fitted to the data. This enables us to use the available data directly for the training of the ML algorithm. Furthermore, ML methods allow for the use of higher dimensional feature vectors such that in future work, information about the material properties and the microstructure of the material can be directly used as input into one single ML yield function able to handle different microstructures.

In this work, Support Vector Classification (SVC) is applied to categorize data sets consisting of principal stresses into the classes "elastic" and "plastic". During training, SVC constructs a hyperplane in stress space, which separates the two regions from each other. Consequently, this hyperplane, defined by the zeros of the so-called SVC decision function, is the equivalent to the yield locus, defined by the zeros of the yield function, and it is constructed such that it has the largest distance to the nearest training data points of both classes. The SVC decision function is defined as [15]

$$f_{\text{SVC}}(\boldsymbol{s}) = \sum_{k=1}^{n} y_k \alpha_k K(\boldsymbol{s}_{\text{sv}}^{(k)}, \boldsymbol{s}) + \rho, \tag{18}$$

where n is the number of support vectors identified during the training process and

$$K(\boldsymbol{s}_{\text{sv}}^{(k)}, \boldsymbol{s}) = \exp\left(-\gamma \|\boldsymbol{s} - \boldsymbol{s}_{\text{sv}}^{(k)}\|^2\right) \tag{19}$$

is the radial basis function (RBF) kernel of the SVC, which is well suited for non-linear problems, with the parameter γ that determines how fast the influence of one support vector decays in stress space. The support vectors $\boldsymbol{s}_{\text{sv}}^{(k)}$, the dual coefficients $y_k \alpha_k$, and the intercept ρ are determined during the training. There are essentially two parameters that control the training process and thus the quality of the obtained decision function: (i) $\gamma > 0$ is a parameter of the RBF kernel function and controls how far-reaching the influence of each support vector is: the larger the value of γ, the more short-ranged and local the influence; (ii) $C > 0$ is a parameter that is used only during the training to regularize the decision function, but that does not directly enter the decision function (18). The larger the value of C, the more flexible but irregular the decision function will become by approximating the shape of the training data more accurately. The choice if these training parameters is critical for the successful use of the decision function in a flow rule. In short, the larger both values are, the more flexible and sensitive to local values the resulting decision function will become. Thus, too small values will result in a smooth but not accurate approximation of the true yield function, whereas too large values will result in a noisy yield function that cannot be used in FEA. The numerical example presented later on will demonstrate this effect and provide examples for values producing accurate yet sufficiently smooth results for the yield function.

For the supervised training, a set of n_t feature vectors $\boldsymbol{s}_{\text{train}}^{(j)}$ has to be provided together with the result vector $y^{(j)}$ with $j = (1, ..., n_t)$, which takes values only in two categories: $y^{(j)} = -1$ for those training data points $\boldsymbol{s}_{\text{train}}^{(j)}$ in the "elastic" regime and $y^{(j)} = +1$ for training data in the "plastic" regime. This training data are within the core of the method outlined here, because during the training process they are directly used to define the support vectors that in turn determine the plastic properties of the material. It is, therefore, essential to have sufficiently many training data points in close proximity to the yield locus to approximate it accurately. However, the SVC training will create support vectors only in the region covered by the training data, and outside this region the decision function drops to zero, which could produce erroneous results if the elastic predictor step of the return mapping algorithm falls into such a region. Hence, data points that lie deeper within the elastic and plastic regions are required to prevent the decision function (18) from falling back to zero. Such data points, however, can be constructed from available raw data lying close to the yield locus simply by linearly scaling principal stresses in the elastic region towards smaller values, such that they stay within the elastic region, and, likewise in the plastic region, scaling principal stress data towards higher values. Thus, the raw data can be spread throughout the stress space, even without knowing the strain value associated with each data point. Only the knowledge of its class "elastic" or "plastic" is required as knowledge for this data extension step improving the training process. This property of the formalism introduced in this work is of critical importance, because it enables the creation of large volumes of training data from relatively few raw data points close to the yield locus, as demonstrated below.

As seen from Equation (18), the decision function is a continuous function constructed in a way to reproduce this category, i.e., the sign of the training data in the respective regions in the optimal way. To make predictions about the elastic or plastic material behavior at any given stress, the sign of the value of the decision function $f_{\text{SVC}}(\boldsymbol{s})$ at the given stress is evaluated. Furthermore, the yield locus of the material can be obtained in the same way as for traditional yield functions, simply by finding the zeros of the continuous function. In this way, furthermore, the distance of any point in stress space to the yield locus can be evaluated, which is important for making efficient predictor steps during FEA and for calculating plastic strain increments for the return mapping algorithm. In particular, as described in the previous section, the gradient to the yield locus needs to be known in order to calculate plastic strain increments, that bring the stress back to the yield surface, because the plastic material does not support any stresses outside. Due to the definition of the ML yield function as convolution sum over the support vectors, the gradient to the SVC decision function can be calculated analytically as

$$\frac{\partial f_{\text{SVC}}}{\partial \boldsymbol{s}} = \sum_{k=1}^{n} y_k \alpha_k \frac{\partial K(\boldsymbol{s}_{\text{sv}}^{(k)}, \boldsymbol{s})}{\partial \boldsymbol{s}} \tag{20}$$

with

$$\frac{\partial K(\boldsymbol{s}_{sv}^{(k)}, \boldsymbol{s})}{\partial \boldsymbol{s}} = -2\gamma \exp\left(-\gamma \|\boldsymbol{s} - \boldsymbol{s}_{\text{sv}}^{(k)}\|^2\right)(\boldsymbol{s} - \boldsymbol{s}_{\text{sv}}^{(k)}). \tag{21}$$

The gradient of the yield function in the 3D principle stress space is obtained by multiplication of the gradient in the cylindrical stress space with the Jacobian defined in Equation (14). Thus, the formulation of the data-oriented yield function based on the SVC algorithm can be used directly as ML yield function in FEA, with the same formalism for plasticity as for standard yield functions.

3. Results

In the following, it will be demonstrated how the derived formulation of the ML yield function can be trained with data obtained from a reference material and used in FEA as constitutive law for plasticity. All numerical examples are conducted with the tools provided on the open-source platform Sci-Kit Learn [22] and a Python library for FEA written by the author of this work. The Python code used for generating the results presented here is provided as supplementary material in form of a Jupyter notebook.

3.1. Training of ML Yield Function

A reference material with Hill-type anisotropy is defined with the material parameters as given in Table 1. The yield locus of this reference material for plane-stress conditions with $\sigma_3 = 0$ is plotted in Figure 1, in which the yield locus of the reference material with Hill-like anisotropy is compared with that of an isotropic material with the same yield strength σ_y.

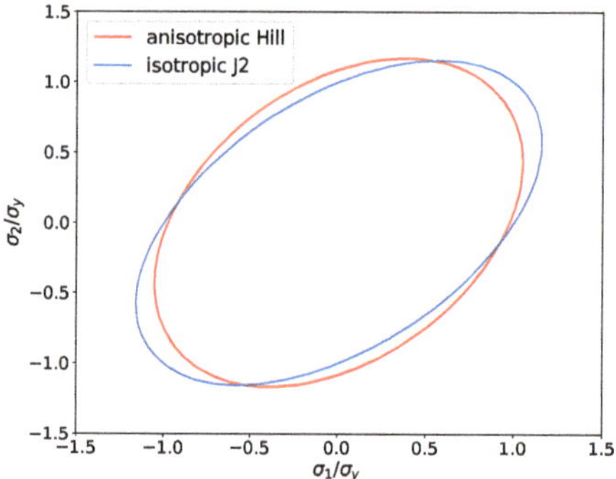

Figure 1. Yield locus for plane-stress conditions ($\sigma_3 = 0$) and Hill-like anisotropy with parameters given in Table 1 (red line) and for an isotropic material with the same yield strength, but $H_1 = H_2 = H_3 = 1$ (blue line). The values of the principal stresses are normalized by the yield strength σ_y.

Table 1. Elastic and plastic material parameters defining the reference material with Hill-like anisotropy in its plastic flow behavior. For simplicity, isotropic elastic behavior and ideal plasticity without work hardening are assumed in this work.

Quantity	Symbol	Value
Young's modulus	E	200 GPa
Poisson's number	ν	0.3
Yield strength	σ_y	150 MPa
Hill parameters	H_1, H_2, H_3	0.7, 1, 1.4

The thus-defined reference material is used to produce training data—and later also test data—for the machine learning algorithm. To accomplish this, a set of stress values in form of principal deviatoric stresses is produced in a way to cover the complete space of polar angles and also sufficiently many equivalent stresses in the elastic and plastic regimes. This is conveniently achieved by creating a set of n_{ang} equally distributed polar angles $\theta^{(k)}$ in the range of $[-\pi, \pi]$ and a set of n_s equivalent stresses $\sigma_{eq}^{J2\,(l)}$ in the range $[0.1\sigma_y, 5\sigma_y]$. Note that for the entire procedure, the yield strength σ_y of the material is assumed to be known. This is not a restriction, because the yield strength of an unknown material can be easily determined from the input data in a pre-analysis step.

The transformation into principal stresses is performed as

$$\sigma_{\text{train}}^{(j)} = \sqrt{2/3}\,\sigma_{eq}^{J2\,(l)} \left(a \cos \theta^{(k)} + b \sin \theta^{(k)} \right) \tag{22}$$

with

$$j = k + (l - 1)n_{\text{ang}} \qquad (k = 1, ..., n_{\text{ang}};\ l = 1, ..., n_s) \tag{23}$$

and the unit angles a and b spanning the deviatoric stress plane as given above. This produces a set of $n_t = n_{\text{ang}}n_s$ principal stresses with which, finally, the set of result vectors

$$y^{(j)} = \text{sgn}\left(f(\sigma_{\text{train}}^{(j)}) \right) \qquad \text{with} \qquad j = (1, ..., n_t), \tag{24}$$

is generated by evaluating the yield function f of the reference material, as defined in Equations (2) and (5), with the material parameters given in Table 1.

In the numerical example given here, the full training data set comprises $n_{\text{ang}} = 36$ values for the polar angle and $n_s = 28$ values for the equivalent stress for each angle, resulting in a total of $n_t = 1008$ training data sets. Concerning the effort to create this data, it is noted here that only the number of angles n_{ang} is relevant for the number of experiments or micromechanical simulations necessary to generate the training data, because each angle defines a load case from which several stresses in the elastic and plastic regime will result, and other training data points can be easily constructed from this raw data by linear scaling, as described above. The implications of the number of load cases required to achieve an accurate representation of the ML yield function will be further discussed in Section 4. A graphical representation of the training data is shown in Figure 2, where also the different ways of representing the anisotropy of the yield function with Hill-type equivalent stresses and von Mises (J2) equivalent stresses are demonstrated. The actual training data comprise four additional sets of polar angles associated with larger equivalent stresses scaled to values of up to $\sigma_{eq} = 5\sigma_y$ to prevent the decision function from falling back to zero in this regime, which might cause erroneous results in FEA. To enforce the periodicity of the training ML yield function and its gradient, the training data is periodically repeated within the training algorithm, such that the polar angle covers a range $-1.3\pi < \theta < 1.3\pi$.

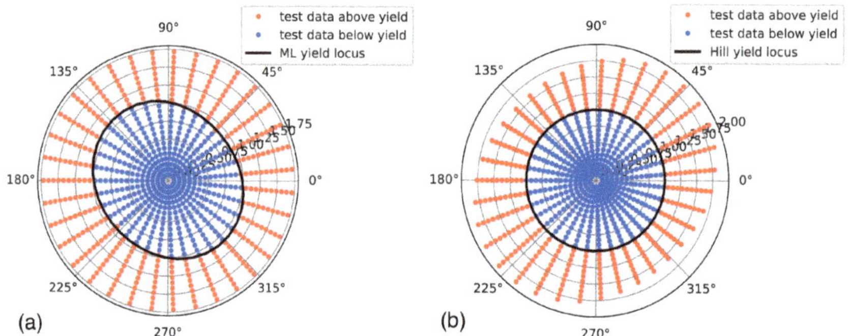

Figure 2. Polar plots of a subset of the training data produced from the anisotropic yield function of the reference material: (**a**) Von Mises (J2) equivalent stresses according to Equation (3) are used, such that the yield strength, rather than the equivalent stress, is a function of the polar angle θ. (**b**) Equivalent stresses are calculated according to the Hill definition in Equation (5) to achieve a constant yield strength by mapping the equivalent stresses accordingly. In both figures, the yield locus is indicated by a solid black line, data points in the elastic regime are plotted in blue color and data in the plastic regime in red color. Both figures represent the same stress data, only mapped in a different way; all stresses are normalized with the reference yield strength σ_y.

With this data set, the training of the SVC algorithm is performed. Using the training parameters $C = 10$ and $\gamma = 4$ results in a very good training score of above 99%. However, to evaluate the true quality of the training procedure and to judge whether overfitting has occurred, it is necessary to verify the results with an independent set of test data, which has not been used for training purposes. The error produced on such test data sets with 480 random deviatoric stresses as data points is below 1%, and the R^2-correlation coefficient between test data and training data is above 98%, which leads to the conclusion that the trained ML yield function has a very high accuracy and robustness. A variation of the training parameters revealed that the results are rather insensitive to the parameter C, which can be varied between $2 < C < 20$ without having a pronounced influence on the results, whereas changing the parameter γ by more then 20% causes a significant deterioration of the training results. The resulting SVC decision function, defined in Equation (18), is plotted together with the training data in Figure 3 in the deviatoric stress space.

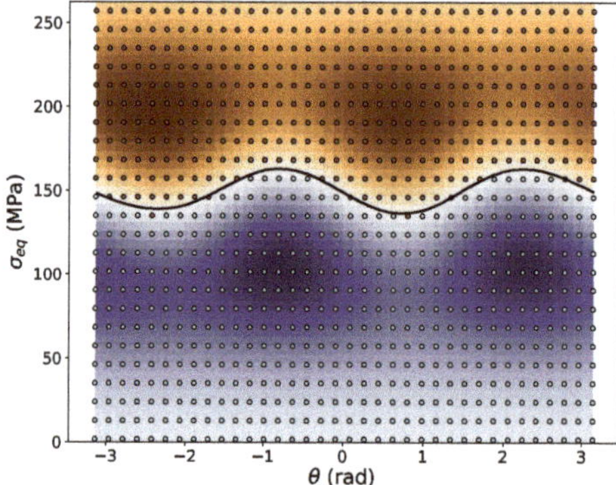

Figure 3. Field plot of the trained SVM decision function defined in Equation (18), where areas in purple color shades represent negative values and brown shades represent positive values. The numerical value of the decision function is not relevant because only its sign is taken into account in the flow rule. The isoline for $f_{SVC} = 0$ is represented as a black line. Training data are plotted in light blue color for data with negative values (elastic) and in brown color for positive values (plastic).

Finally, to demonstrate the quality of the ML yield function in the full principal stress space, the predicted categories are plotted in three different slices corresponding to different plane-stress conditions, together with the yield function of the reference material in Figure 4. It is seen that the training data in the deviatoric space covers in fact only a single line in each slice, which demonstrates the power of reducing the dimensionality of the data-oriented yield function by exploiting basic physical principles.

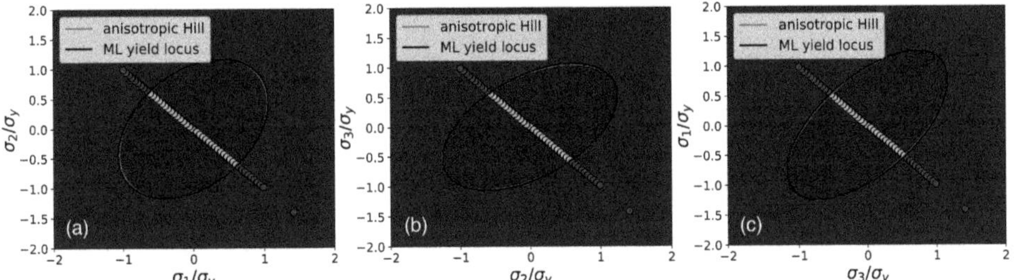

Figure 4. Color map of the trained SVC prediction of the yield function in slices through the principal stress space defined by plane-stress conditions: (**a**) $\sigma_3 = 0$; (**b**) $\sigma_1 = 0$; (**c**) $\sigma_2 = 0$. Brown regions indicate values of "+1" (plasticity) and purple regions values of "-1" (elasticity). The ML yield locus, corresponding to the isoline for $f_{SVC} = 0$, is represented as a black line; the yield locus of the Hill-like anisotropic reference material is indicated as a red line. The training data points are plotted with the same color code as in Figure 3.

3.2. Application of The Trained ML Yield Function in FE Analysis

The ML yield function trained and analyzed in the previous step shall now be applied in FEA to demonstrate its usefulness for this purpose. The numerical examples provided here have been conducted with the Python library "pyLab-FE" created by the author, which is provided in the supplementary materials together with a Jupyter notebook following the work-flow defined in this work. The known parameters and support vectors resulting from the training process of the ML yield function together with the mathematical formalism laid out in Section 2 allow a rather straightforward evaluation of the yield function as sum over the support vectors convoluted with the kernel function, such that they can also be used for implementing a user material subroutine (UMAT) for common commercial FEA tools in any compiler language.

As numerical examples, four different load cases are simulated with FEA: (i) uniaxial stress in horizontal direction, (ii) uniaxial stress in vertical direction, (iii) equibiaxial strain under plane-stress conditions, and (iv) pure shear strain under plane-stress conditions. The simple finite element model used to study these load cases consists of four quadrilateral elements with linear shape functions and full integration, as shown in Figure 5. For all load cases, plane stress conditions with $\sigma_3 = 0$ are enforced and the normal degrees of freedom (dof) for the boundary nodes are prescribed, while all boundary nodes are allowed to relax to their equilibrium positions along the boundary. The bottom and the left-hand-side nodes are always restricted to a normal displacement of zero. For the uniaxial load cases, tensile displacements are prescribed either on the top or on the right-hand-side (rhs) nodes, while the other boundary is force-free, resulting in a uniaxial stress. For equibiaxial strain, the top and the rhs boundary nodes are subjected to identical displacements; whereas for pure shear, the rhs nodes are loaded with the negative displacement applied on the top nodes. By virtue of these boundary conditions, the deformation causes only normal stresses and strains, but no shear components. Hence, the restrictions of the formulation of the ML flow rule are fulfilled, and the material axes remain aligned with the directions of the principal stresses.

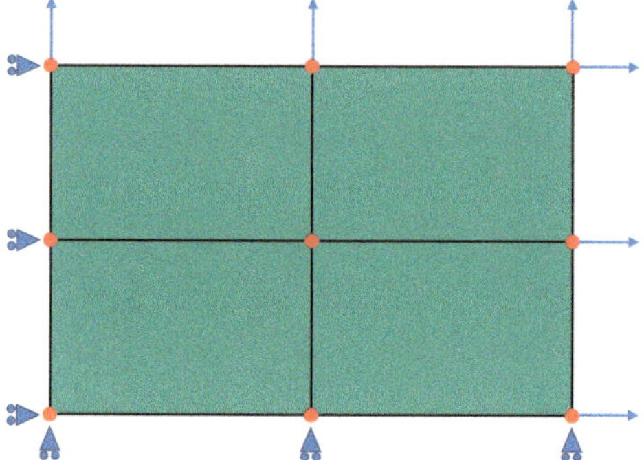

Figure 5. The finite element model on which four different load cases are studied consists of four quadrilateral elements (green) with linear shape function, and at total of nine nodes (red) situated at the corners of the elements. The bottom and left-hand-side boundary nodes are restricted to zero normal displacement (blue triangles), and the loading is applied on top and right-hand-side-nodes (blue arrows), as described in the text.

These four load cases are applied to the reference material as well as to the material with the ML yield function, and the resulting yield stresses and plastic strains at the end of each load step are compared in Table 2. In Figure 6, the resulting global equivalent stresses and equivalent total strains

$$\epsilon_{eq} = \sqrt{\boldsymbol{\epsilon} : \boldsymbol{\epsilon}} \tag{25}$$

for each load case are plotted for both materials, where the different definitions of the equivalent stress have been applied to the reference material.

Table 2. Yield stresses (YS) obtained for Hill-like yield function, with parameters given in Table 1, and machine learning (ML) yield function under the specified load cases. The relative errors in yield stress and equivalent plastic strain (PE) at maximum load are also specified.

Load Case	YS-Hill (MPa)	YS-ML (MPa)	Rel. Error YS	Rel. Error PE
uniaxial stress, horizontal	146.4	148.4	1.41%	-1.95%
uniaxial stress, vertical	162.7	162.2	-0.3%	-0.45%
equibiaxial strain	136.9	139.5	1.88%	-1.36%
pure shear strain	161.1	159.7	-0.87%	0.24%

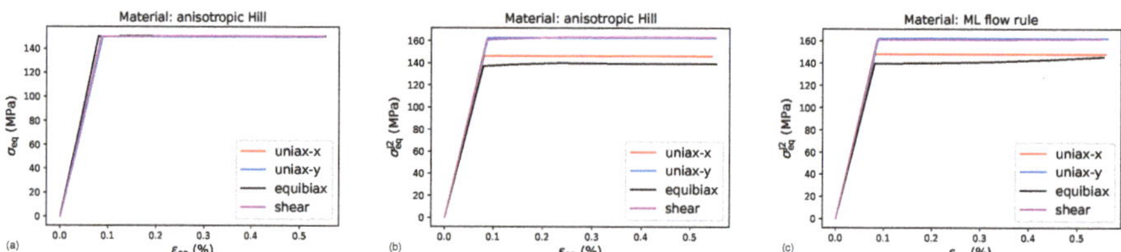

Figure 6. Stress strain curves obtained for elastic-ideal plastic material behavior under the loading conditions specified in the legend: (**a**) Equivalent total strain vs. equivalent Hill-stress, (**b**) equivalent total strain vs. equivalent J2-stress for Hill-like yield function, and (**c**) equivalent total strain vs. equivalent J2-stress for ML yield function.

To further demonstrate the correctness of the plastic behavior resulting from the ML flow rule, the flow stresses of the material, i.e., the stresses occurring during plastic deformation, are plotted together with the yield locus. For ideal plasticity, treated in this work, it is expected that the flow stress remains on the yield locus, since the material does not sustain larger stresses. In Figure 7 it is shown that this expectation is fulfilled to a very good degree, by comparing the yield loci and the element solutions of the flow stresses obtained for the ML yield function with those of the reference material with a Hill-like flow rule.

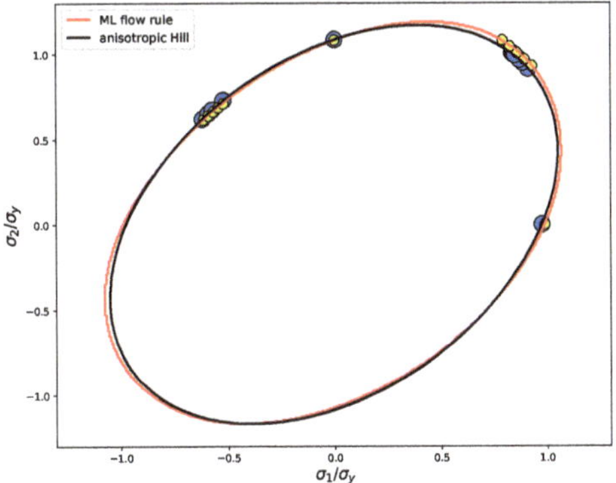

Figure 7. Stress states obtained for the four different plane-stress load cases are plotted in the σ_1-σ_2 plane together with the yield loci of the trained ML flow rule and the Hill-like reference material. The flow stresses resulting from the ML yield function are plotted as small yellow circles and those from anisotropic Hill plasticity as large blue circles.

3.3. Tresca Flow Rule

In the next example, the ML yield function is trained with data from a reference material with a Tresca yield function, which is fundamentally different from the elliptical Hill-like yield functions. The Tresca equivalent stress is defined as

$$\sigma_{eq}^{\text{Tresca}} = \sigma_{\text{I}} - \sigma_{\text{III}}, \tag{26}$$

where σ_{I} is the largest principle stress and σ_{III} is the smallest principle stress [23]. Using the Tresca equivalent stress in the yield function (2) leads to an isotropic plastic deformation of the material, however, with very different characteristics than for a J2 equivalent stress. Hence, it is a critical test for the new method developed in this work to apply it to such yield functions. The training data for this yield function has been produced in the same way as before. However, with $n_{\text{ang}} = 600$ data points for the polar angle, a much larger number of training data points has been required to follow the subtle features of the Tresca flow rule. Furthermore, the training parameters $C = 50$ and $\gamma = 9$ have been applied to allow for sufficient flexibility of the ML yield function to approximate the abrupt changes of the yield behavior in the deviatoric stress space, *cf.* Figure 8. The scores with test and training data are above 99%, and the R^2-value on test data is 96%. Even with this training procedure, the sharp corners of the Tresca yield locus are slightly rounded off by the ML yield function, leading to somewhat lower yield strengths of the material in these directions, as seen in Figure 9a, where the resulting stress-strain curves are plotted as J2 equivalent stress vs. equivalent total strain. In Figure 9b the yield loci of the Tresca reference material, the trained ML yield function and an isotropic J2 material are plotted in comparison. Furthermore, the flow stresses resulting from the FEA are plotted in this graph to verify that they lie on the ML yield locus, as expected for ideal plasticity.

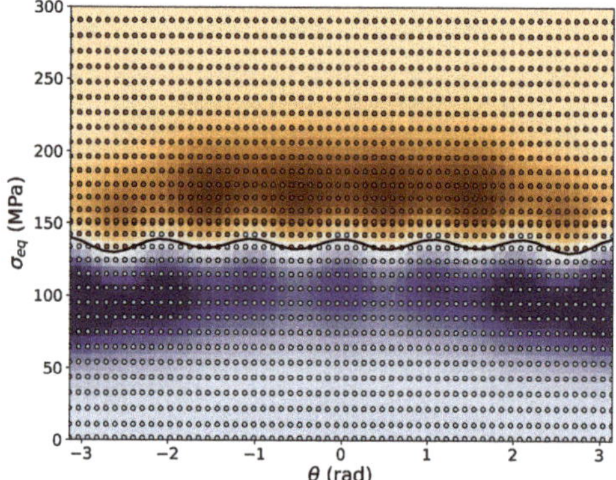

Figure 8. Field plot of the ML yield function trained with data generated from a reference material with a Tresca yield criterion, where areas in purple color shades represent negative values and brown shades represent positive values. The isoline, where the ML yield function is zero, is plotted as black line. The test data points are plotted as brown circles, for stresses in the plastic regime, and a s light blue circles for stresses in the elastic regime.

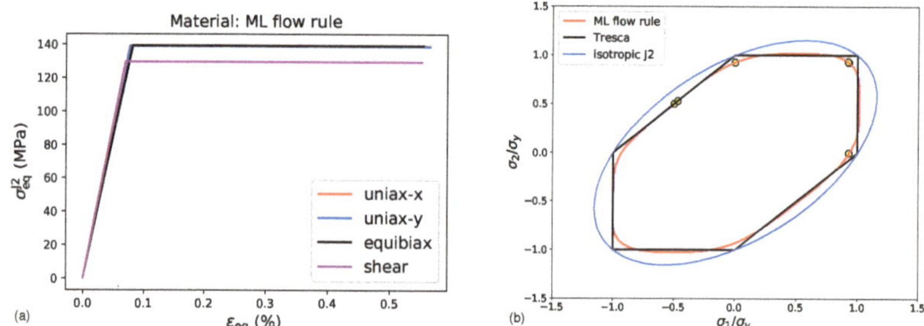

Figure 9. Results of FEA on the material with an ML flow rule trained with data obtained from a Tresca yield criterion: (**a**) Equivalent J2 stress plotted over the equivalent total strain, for the four different load cases given in Table 2. (**b**) Flow stresses obtained with the ML yield function plotted as yellow circles in the σ_1-σ_2 principle stress space, together with the yield loci of the ML flow rule, the Tresca flow rule and an isotropic J2 flow rule.

With these numerical examples, the applicability of the ML yield function developed in this work has been demonstrated for two fundamentally different kinds of flow behavior: anisotropic Hill-like plasticity and isotropic Tresca plasticity. The new formulation has proven to be numerically stable. The numerical effort is somewhat higher than that for mathematical closed-form yield functions, because the calculation of the predictor step requires a higher effort in numerically evaluating the distance of a given point in stress space to the yield locus. However, in conjunction with the implementation of the ML yield function in a compiled computer code, FEA even for large engineering models seems to be feasible with the new ML yield functions.

4. Discussion

In this section, the requirements on training data for the ML yield function will be more closely examined. For the first application of the ML yield function, training data for $n_{\mathrm{ang}} = 36$ load cases have been constructed from a reference material with Hill-like anisotropy, and it has been demonstrated that this number of load cases produces rather accurate results in this case, whereas the approximation of a Tresca yield function required $n_{\mathrm{ang}} = 600$ load cases. Numerical studies with larger data sets reveal that for Hill-like anisotropic yield functions the accuracy of the results increases slightly for data sets of up to $n_{\mathrm{ang}} = 200$. The accuracy as well as the numerical stability and efficiency of the method remain stable for even larger data sets, which has been tested for up to $n_{\mathrm{ang}} = 600$. It is also interesting to see that the accuracy of the method for Hill-like anisotropic yield functions is only slightly reduced when rather small data sets of $n_{\mathrm{ang}} = 8$ load cases are used. Even producing the training data under plane-stress conditions with $\sigma_3 = 0$ does not change the quality of the results significantly, which is a consequence of mapping all stresses onto the deviatoric plane and extending the results by assuming that the material's flow behavior does not depend on hydrostatic stress components, which is fulfilled to a very good degree for metals. If it is assumed, furthermore, that the material under consideration shows a symmetric flow behavior under tension and compression, only one half-space of the deviatoric plane needs to be characterized, and the results can be mirrored into the other half-plane. Thus, only the four load cases given in Table 2, have been sufficient to produce training data from which a useful ML yield function for Hill-like anisotropic plasticity results. This example is also provided in the supplementary material.

From these considerations, it can be concluded that for material with an anisotropic flow behavior that can be described with a Hill-like formulation, a small number of experiments under plane-stress conditions, as they can be performed on a bi-axial test rig, are fully sufficient to produce enough raw data to train the ML yield function. Of course, this experimental data could also be used to calculate the Hill parameters. However, the ML yield function offers a larger flexibility, and the training process for machine learning methods like support vector classification (SVC) is highly efficient.

Materials with a more irregularly shaped yield function, like the Tresca yield function, pose much higher demands on the available data and, furthermore, the parameters for the training of the SVC algorithm need to be adapted to allow for more flexibility. The obtained yield function still rounds off the sharp corners of the Tresca yield function such that the result resembles yield functions that can also be obtained with the 18-parameter model of Barlat et al. [4]. Again, it would be possible to determine these parameters directly from the available data, however, with an even higher effort than for the Hill parameters. Yet, the training effort for the ML yield function remains the same, independent of the volume of training data.

In general, it has been verified that only knowing the full stress tensors at the onset of plastic yielding is sufficient to train the ML yield function, and further information, e.g., on plastic strains, is not required. The only relevant information that needs to be known is whether for a given stress state the material response is purely elastic or elastic-plastic. Concerning the best strategy to produce training data, the raw data should lie in close proximity to the yield surface to cover the onset of yielding in an accurate way. Since support vectors are only produced from training data closest to the hyperplane separating the two categories "elastic" and "plastic", however, using only this raw data would cause the SVC decision function (18) to drop to zero quickly outside the region covered by the support vectors. Hence, it is important to downscale the "elastic" raw data points and to upscale the "plastic" raw data points to finally cover the entire relevant region of the deviatoric stress plane and to produce sufficiently many support vectors in this domain. Comparing Figure 3 (Hill-like yield function) and Figure 8 (Tresca yield function), the influence of the number of training points is quite clearly seen, as in the former case 145 support vectors are created during the training process, whereas in the latter case 1689 support vectors are necessary to describe the sharp corners of the Tresca yield function. This example also demonstrates how the number of 36 vs. 600 raw data points is reflected in the number of generated support vectors.

Data-oriented constitutive modeling thus requires only a rather limited amount of data, as compared to other approaches in data-driven mechanics [9,11,13]. Large data volumes, of course, help to increase the accuracy of the resulting ML yield function, but the results achieved even with small amounts of data provide already a very good estimate of the material's anisotropic flow behavior even for load cases that have not been tested. These comparatively moderate requirements on the volume of training data are a consequence of exploiting physical symmetry conditions on the material's flow behavior in the formulation of the data-oriented yield function. For materials exhibiting a significant influence of the hydrostatic stress component on the plastic behavior, the method is still applicable, but the requirements on the training data will be higher.

Another aspect to be discussed here is the use of micromechanical models to produce training data. With such models, the mechanical behavior of realistic microstructures can be simulated with crystal plasticity methods [7], providing an accurate description of the plastic properties of polycrystaline metals with different microstructures and crystallographic textures. One disadvantage of such micromechanical models is their tremendously high numerical effort making them unsuited for FEA applications of engineering structures that are much larger than the grain size of a material. However, it is possible to employ relatively small micromechanical models, validated by experimental data, for creating a sufficiently large data volume describing the mechanical properties of the real material under various loading conditions with a high accuracy. By purposefully varying the microstructure or the texture of the model material, micromechanical simulations also serve the purpose of extending experimental data. This hybrid experimental and numerical data can then be used for the training of the ML yield function presented in this work. In this way, material parameters like grain size, grain morphology, and crystallographic texture can be explicitly included into the feature vector of the ML yield functions, in addition to the purely mechanical data used currently as input for the yield function. This microstructure-sensitive ML yield function can then be used in large-scale FEA for the simulation of engineering structures, which holds the possibility to consider the trained ML flow rule as a "digital twin" of the material, containing all relevant information on the material properties. The data-oriented constitutive model developed in this work will, hence, also pave the way for new approaches to scale-bridging materials modeling.

Critical issues that remain to be solved before the new method can be applied generally in FEA, include (i) the augmentation of the formulation with respect to shear components of stress and strain, (ii) regularization of the ML yield function to ensure its convexity and to the reduce noise in its gradient, and (iii) a data-oriented formulation of work-hardening. Concerning the latter point, it is noted that the current formulation allows the use of the standard methods of isotropic and kinematic work hardening if the hardening parameters are known, because it already contains the gradient of the yield function to calculate the plastic strain rate and also the tangent stiffness. However, a data-oriented formulation of work hardening, e.g., following the ideas of Chinesta et al. [13], would be more consistent with the idea of data-oriented constitutive modeling.

5. Conclusions

In this work, a new formulation of a data-oriented constitutive model for plasticity has been derived and applied within finite element analysis. The central element of this new constitutive model is a support vector characterization (SVC) algorithm serving as yield function. This SVC algorithm is trained by using deviatoric stresses as input data and the information whether a given stress state leads to purely elastic or rather to elastic-plastic deformation of the material as result data. In this way, a machine learning (ML) yield function is obtained, which can determine whether a given stress state lies inside or outside of the elastic regime of the material. Furthermore, the yield locus, i.e., the hyperplane in stress space on which plastic deformation occurs, can be reconstructed from the SVC, and the gradient on this yield locus can be conveniently calculated. Therefore, the standard formulations of continuum plasticity, as the return mapping algorithm, can be applied in finite element analysis in the usual way. Thus, it has been demonstrated that the new ML yield function can replace conventional yield functions in finite element analysis. The main advantage of such data-oriented constitutive models over the conventional ones is that they can be used with higher-dimensional feature vectors combining mechanical stresses with microstructural parameters of a material. In forthcoming work, it will thus be demonstrated how a single ML yield function can be trained to be used as a constitutive rule for a material in different microstructural states. The production of training data by micromechanical models, based on crystal plasticity and a discrete representation of the material's microstructure, allows the ML flow rule to serve as efficient homogenization scheme, which offers new possibilities in scale-bridging material modeling.

Supplementary Materials: Supporting material in the form of a Python library for finite element analysis and a Jupyter notebook with the codes that have been used to produce the results presented in this work are available online at http://www.mdpi.com/1996-1944/13/7/1600/s1 and as a public repository on https://github.com/AHartmaier/pyLabFEA.git under the GNU General Public License v3.0.

Funding: This work has been funded by the Deutsche Forschungsgemeinschaft (DFG, German Research Foundation)—Project-ID 190389738—TRR 103.

Acknowledgments: Helpful discussion with Tobias Glasmachers of Ruhr-Universität Bochum are gratefully acknowledged. Machine learning algorithms have been adopted from the scikit-learn platform (https://scikit-learn.org/stable/).

Conflicts of Interest: The author declares no conflict of interest.

References

1. De Borst, R.; Crisfield, M.A.; Remmers, J.J.; Verhoosel, C.V. *NOnlinear Finite Element Analysis of Solids and Structures*; John Wiley & Sons: Hoboken, NJ, USA, 2012.
2. Hill, R. A variational principle of maximum plastic work in classical plasticity. *Q. J. Mech. Appl. Math.* **1948**, *1*, 18–28. [CrossRef]
3. Hill, R. Constitutive modelling of orthotropic plasticity in sheet metals. *J. Mech. Phys. Solids* **1990**, *38*, 405–417. [CrossRef]
4. Barlat, F.; Aretz, H.; Yoon, J.W.; Karabin, M.; Brem, J.; Dick, R. Linear transfomation-based anisotropic yield functions. *Int. J. Plast.* **2005**, *21*, 1009–1039. [CrossRef]
5. Soyarslan, C.; Klusemann, B.; Bargmann, S. The effect of yield surface curvature change by cross hardening on forming limit diagrams of sheets. *Int. J. Mech. Sci.* **2016**, *117*, 53–66. [CrossRef]
6. Güner, A.; Soyarslan, C.; Brosius, A.; Tekkaya, A. Characterization of anisotropy of sheet metals employing inhomogeneous strain fields for Yld2000-2D yield function. *Int. J. Solids Struct.* **2012**, *49*, 3517–3527. [CrossRef]
7. Vajragupta, N.; Ahmed, S.; Boeff, M.; Ma, A.; Hartmaier, A. Micromechanical modeling approach to derive the yield surface for BCC and FCC steels using statistically informed microstructure models and nonlocal crystal plasticity. *Phys. Mesomech.* **2017**, *20*, 343–352. [CrossRef]
8. Biswas, A.; Prasad, M.R.; Vajragupta, N.; Hassan, H.U.; Brenne, F.; Niendorf, T.; Hartmaier, A. Influence of Microstructural Features on the Strain Hardening Behavior of Additively Manufactured Metallic Components. *Adv. Eng. Mater.* **2019**, *21*, 1900275. [CrossRef]
9. Kirchdoerfer, T.; Ortiz, M. Data-driven computational mechanics. *Comput. Methods Appl. Mech. Eng.* **2016**, *304*, 81–101. [CrossRef]
10. Kirchdoerfer, T.; Ortiz, M. Data-driven computing in dynamics. *Int. J. Numer. Methods Eng.* **2018**, *113*, 1697–1710. [CrossRef]
11. Eggersmann, R.; Kirchdoerfer, T.; Reese, S.; Stainier, L.; Ortiz, M. Model-free data-driven inelasticity. *Comput. Methods Appl. Mech. Eng.* **2019**, *350*, 81–99. [CrossRef]
12. Vorkov, V.; García, A.T.; Rodrigues, G.C.; Duflou, J.R. Data-driven prediction of air bending. *Procedia Manuf.* **2019**, *29*, 177–184. [CrossRef]
13. Chinesta, F.; Ladeveze, P.; Ibanez, R.; Aguado, J.V.; Abisset-Chavanne, E.; Cueto, E. Data-driven computational plasticity. *Procedia Eng.* **2017**, *207*, 209–214. [CrossRef]
14. Bock, F.E.; Aydin, R.C.; Cyron, C.J.; Huber, N.; Kalidindi, S.R.; Klusemann, B. A Review of the Application of Machine Learning and Data Mining Approaches in Continuum Materials Mechanics. *Front. Mater.* **2019**, *6*, 110. [CrossRef]
15. Suykens, J.A.; Vandewalle, J. Least squares support vector machine classifiers. *Neural Process. Lett.* **1999**, *9*, 293–300. [CrossRef]
16. Vapnik, V. The support vector method of function estimation. In *Nonlinear Modeling*; Springer: Berlin/Heidelberg, Germany, 1998; pp. 55–85.

17. Reimann, D.; Chandra, K.; Vajragupta, N.; Glasmachers, T.; Junker, P.; Hartmaier, A. Modeling macroscopic material behavior with machine learning algorithms trained by micromechanical simulations. *Front. Mater.* **2019**, *6*, 181. [CrossRef]

18. Gunn, S.R. Support vector machines for classification and regression. *ISIS Tech. Rep.* **1998**, *14*, 5–16.

19. Hastie, T.; Tibshirani, R.; Friedman, J. *The Elements of Statistical Learning: Data Mining, Inference, and Prediction*; Springer Science & Business Media: Berlin/Heidelberg, Germany, 2009.

20. Mises, R.V. Mechanics of solid bodies in the plastically-deformable state. *Göttin Nachr. Math. Phys.* **1913**, *1*, 582–592.

21. Lode, W. Versuche über den Einfluß der mittleren Hauptspannung auf das Fließen der Metalle Eisen, Kupfer und Nickel. *Z. Phys.* **1926**, *36*, 913–939. [CrossRef]

22. Pedregosa, F.; Varoquaux, G.; Gramfort, A.; Michel, V.; Thirion, B.; Grisel, O.; Blondel, M.; Prettenhofer, P.; Weiss, R.; Dubourg, V.; et al. Scikit-learn: Machine Learning in Python. *J. Mach. Learn. Res.* **2011**, *12*, 2825–2830.

23. Tresca, M.H. On further applications of the flow of solids. *Proc. Inst. Mech. Eng.* **1878**, *29*, 301–345. [CrossRef]

 *materials*

Article

A Stochastic FE^2 Data-Driven Method for Nonlinear Multiscale Modeling

Xiaoxin Lu [1], Julien Yvonnet [2,*], Leonidas Papadopoulos [3], Ioannis Kalogeris [3] and Vissarion Papadopoulos [3]

1. Shenzhen Institute of advanced electronic materials, Shenzhen Institutes of Advanced Technology, Chinese Academy of Sciences, Shenzhen 518103, China; luxiaoxin.cassie@gmail.com
2. MSME, University Gustave Eiffel, CNRS UMR 8208, F-77454 Marne-la-Vallée, France
3. Department of Civil Engineering, National Technical University of Athens, 15780 Athens, Greece; lew.papado@hotmail.com (L.P.); yianniskalogeris@gmail.com (I.K.); vissarion.papadopoulos@gmail.com (V.P.)
* Correspondence: julien.yvonnet@univ-eiffel.fr

Abstract: A stochastic data-driven multilevel finite-element (FE^2) method is introduced for random nonlinear multiscale calculations. A hybrid neural-network–interpolation (NN–I) scheme is proposed to construct a surrogate model of the macroscopic nonlinear constitutive law from representative-volume-element calculations, whose results are used as input data. Then, a FE^2 method replacing the nonlinear multiscale calculations by the NN–I is developed. The NN–I scheme improved the accuracy of the neural-network surrogate model when insufficient data were available. Due to the achieved reduction in computational time, which was several orders of magnitude less than that to direct FE^2, the use of such a machine-learning method is demonstrated for performing Monte Carlo simulations in nonlinear heterogeneous structures and propagating uncertainties in this context, and the identification of probabilistic models at the macroscale on some quantities of interest. Applications to nonlinear electric conduction in graphene–polymer composites are presented.

Keywords: data-driven; multiscale; nonlinear; stochastics; neural networks

check for **updates**

Citation: Lu, X.; Yvonnet, J.; Papadopoulos, L.; Kalogeris, I., Papadopoulos, V. A Stochastic FE^2 Data-Driven Method for Nonlinear Multiscale Modeling. *Materials* **2021**, 14, 2875. https://doi.org/10.3390/ma14112875

Received: 30 March 2021
Accepted: 21 May 2021
Published: 27 May 2021

Publisher's Note: MDPI stays neutral with regard to jurisdictional claims in published maps and institutional affiliations.

1. Introduction

Predicting the nonlinear behavior of materials from knowledge of their microstructure is a critical topic in engineering. For example, the development of 3D-printed micromaterials [1–3] or of nanomaterials [4,5] with nonlinear behaviors opens exciting opportunities for designing innovative functionalized and enhanced engineering systems. While linear effective properties of heterogeneous materials can be accurately estimated though either analytical [6,7] or numerical techniques [8], predicting the effective behavior of nonlinear materials requires more advanced techniques.

A direct but limited approach is the use of the representative volume element (RVE) to calibrate an empirical nonlinear model. A limitation of such techniques is the number of parameters to be calibrated for complex, nonlinear, or multiphysics problems. To more accurately describe the behavior of general nonlinear materials, the so-called multilevel finite-element (FE^2) method [9–16] or computational homogenization has been developed in recent years. In this approach, an RVE is associated to each Gaussian point of a finite-element macrostructure, and a nonlinear problem must be solved in each integration point and for each iteration of the macrosolving procedure. The drawback of this method, however, is that it induces unaffordable computational times in practical applications.

Several strategies were developed recently to alleviate FE^2 calculations. First, the strategy relies on reducing micro-RVE computations through efficient techniques such as model-order reduction [17,18], fast Fourier transform [19,20], wavelet transforms [21], NTFA [22], self-clustering analysis (SCA) [23,24], or GPU acceleration [25]. In [26], He et al. developed an adaptive strategy to reduce microcalculations by constructing the reduced basis on the fly during the macroscale calculation.

Another idea, initiated in [27,28], is the use of so-called data-driven approaches in which microscale calculations are performed offline, and are then used as data in an online stage to reconstruct the macroscopic (effective) behavior. For this purpose, several techniques were proposed, including interpolation methods [27,29], neural networks [28,30–35], Bayesian inference [36], Fourier series expansions [37], or Gaussian process regression [38]. In the related techniques, offline data collection is used in a regression process to construct an accurate surrogate model whose evaluation is several orders of magnitude lower than that performing one RVE nonlinear calculation. A critical comparison of several regression techniques used in data-driven multiscale approaches can be found in [39]. In [40], Avery et al. investigated and discussed several regression methods with ANN in homogenization problems of hyperelastic woven composites, and demonstrate its use in advanced dynamic or fluid structure applications. Recent advances of data-driven techniques, including handling history-dependent behaviors such as plasticity, can be found in [35,41,42]. On-the-fly construction of the surrogate model by probabilistic machine learning was proposed in [38]. Developments of neural-network techniques in FE^2, including feed-forward and recurrent neural networks, can be found in [31,41]. In [43,44], a manifold-based nonlinear reduced-order model in tandem with a digital database was developed for the nonlinear multiscale analysis of hyperelastic structures involving neural networks, a kernel inverse/reconstruction map, and dimension reduction through an isomap.

Stochastic extensions of data-driven methods in multiscale applications are relatively new and unexplored. One of the first analyses in this context can be found in [45,46], where the NEXP method [27] was extended to stochastic problems. In these studies, stochastic parameters were introduced within the surrogate model using a separated representation-interpolation technique. Probability density functions related to the nonlinear macroscale problem were identified. In [47], a machine-learning strategy based on a three-dimensional convolutional neural network was introduced to evaluate the linear effective properties of random materials from geometrical descriptions of RVE. In [24], a framework for uncertainty quantification in a data-driven approach was proposed where self-consistent clustering analysis (SCA) [23,24] was used to reduce computational times in the learning step.

In this paper, the use of data-driven methods for heterogeneous nonlinear materials with uncertainties at both the micro- and the macroscale is addressed. Taking into account uncertainties in nonlinear multiscale methods implies (a) constructing a probabilistic surrogate macromodel from microcalculations, allowing for generating realizations of the macroresponse for a given macroloading; and (b) performing Monte Carlo simulations of the model at the macroscale to quantify uncertainties on the quantities of interest in the structure. In view of its immense computational requirements, direct use of FE^2 for stochastic nonlinear two-scale analysis is not possible. However, data-driven FE^2 approaches have comparable computational costs as compared to classical (one-scale) FEM calculations, and they open the route to developing stochastic two-scale nonlinear approaches. To the best of our knowledge, this problem remains relatively unexplored in the literature. A new stochastic data-driven approach based on RVE calculations was developed for taking into account random effects in nonlinear heterogeneous structures. First, preliminary RVE calculations were performed. These calculations include several microstructural features that varied, such as the distribution of heterogeneities and its volume fraction. Then, for each realization of the random microstructure, the space of macroscopic loading was sampled, and boundary conditions were prescribed on the RVE. Subsequently, the nonlinear problem was solved by FEM. This large database was used to construct a surrogate model whose inputs were the macroloading and the volume fraction, and its output was the macroscopic (homogenized) response. A new hybrid neural-network–interpolation (NN–I) surrogate model is proposed to provide an accurate response with a limited number of realizations. Once constructed, this model can be used within stochastic analysis of two-scale nonlinear structure calculations. At the macroscale, the volume fraction of heterogeneities is considered to be random here, and it was modeled as a stochastic field with given probabilistic characteristics. Then, during the macro-non-linear resolution, solving the full nonlinear RVE was replaced by the proposed fast surrogate model, which allowed for performing hundreds of macro-non-linear calculations at the cost of classical FEM problems. As a result, statistical postprocessing can be performed on the macroquantities of interest, and probabilistic models could be identified.

The novelties of this paper are twofold. The first is the proposed neural-network–Interpolation FE^2 method, which is an extension of our previous neural-network FE2 method, developed in [28,30]. The NN–I scheme allows for modeling the stochastic spatial variability of the volume fraction in the frame of the FE^2 procedure, leading to the improved accuracy of the surrogate model when limited data are available. The second novelty is the application of this machine-learning method to nonlinear multiscale stochastic problems. Using the proposed approach, FE^2 calculations can be reduced by several orders of magnitude, allowing for Monte Carlo simulation on stochastic nonlinear multiscale structures. It is demonstrated for the first time that uncertainties can be propagated in this context, and probabilistic models can be identified.

The paper is organized as follows. Section 3 presents the equations of the nonlinear RVE problem, and the definitions of the input (macroelectric load) and output (homogenized electric flux) in the nonlinear composite. Section 4 introduces the hybrid neural network/interpolation scheme, and its construction using offline data on RVE is described. In Section 5, the present stochastic data-driven strategy is proposed. Lastly, numerical examples are presented in Section 6.

2. Brief Review of FE^2 Method for Nonlinear Conduction

The multilevel finite-element method [9,10], also called FE^2 in the literature, as it involves two levels of finite-element simulations, and independently proposed by several other authors and groups [11–16], was introduced as a general multiscale method for solving nonlinear heterogeneous structural problems. The basic underlying idea is that two levels of finite elements must be concurrently solved, one for each scale. At the macroscale, each integration point of the finite-element mesh is associated with a representative volume element (RVE). Boundary conditions depending on the macroscopic state (strain, electric field, etc.) are prescribed on the boundary of each RVE. After solving each nonlinear problem at each integration point, the appropriate macroscopic response (stress, electric flux), is averaged over the RVE and provided at the macrointegration point. Then, the macroscopic constitutive law is available only through solving a nonlinear problem. These operations are repeated until convergence is reached at both scales (see Figure 1).

For the sake of simplicity, a brief review of the method in a context of nonlinear conduction is presented. We consider a macroscopic structure associated with a domain $\overline{\Omega} \subset \mathbb{R}^3$, with a boundary $\partial\overline{\Omega}$. The assumption of scale separation is adopted (an extension of the method to second-order homogenization can be found in [14]). The microstructure was assumed to be characterized by an RVE associated with a domain $\Omega \subset \mathbb{R}^3$, with boundary $\partial\Omega$.

In the context of nonlinear electric conduction, electric field $\mathbf{E}(\mathbf{x})$ is related to the electric flux, or electric displacement $\mathbf{j}(\mathbf{x})$ by a nonlinear local constitutive relationship. Electric field $\mathbf{E}$ is defined by $\mathbf{E}(\mathbf{x}) = -\nabla\phi(\mathbf{x})$, where ϕ is the electric potential, $\nabla(.)$ is the gradient operator, and $\mathbf{x}$ is a material point within Ω. In the following, $\overline{(.)}$ notations denote macroscale quantities. For a given macroscopic electric field $\overline{\mathbf{E}}$, the RVE problem is to find $\phi(\mathbf{x})$, such that

$$\nabla \cdot \mathbf{j}(\mathbf{x}) = 0 \ \forall \mathbf{x} \in \Omega, \tag{1}$$

where $\nabla \cdot (.)$ is the divergence operator. The constitutive law is given by

$$\mathbf{j}(\mathbf{x}) = \mathcal{F}^{nl}(\mathbf{E}(\mathbf{x})). \tag{2}$$

where $\mathcal{F}^{nl}$ is a local nonlinear operator (specified in Section 3). The equilibrated electric field should satisfy

$$\overline{\mathbf{E}} = \frac{1}{V} \int_{\Omega} \mathbf{E}(\mathbf{x})d\Omega, \tag{3}$$

where V is the volume of Ω. Equation (3) can be verified, e.g., by the following boundary condition:

$$\phi(\mathbf{x}) = -\overline{\mathbf{E}} \cdot \mathbf{x} + \tilde{\phi}(\mathbf{x}) \quad \text{on } \partial\Omega, \tag{4}$$

where $\tilde{\phi}(\mathbf{x})$ is a periodic function over Ω.

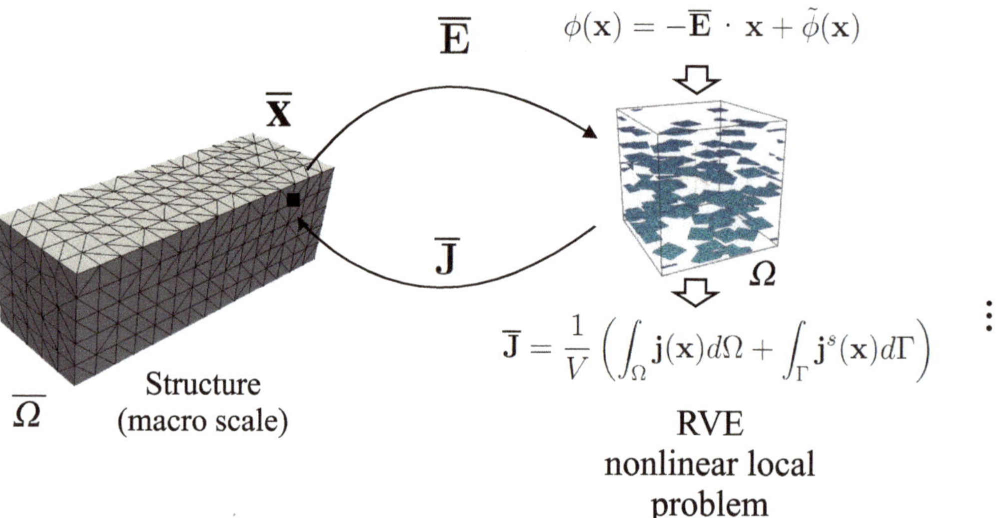

Figure 1. Schematic of classical FE^2 method for nonlinear heterogeneous conduction problem (adapted from [8]).

In the presence of imperfect interfaces and surface electric flux along interfaces (see [48]), the effective electric current $\overline{\mathbf{J}}$ is defined according to

$$\overline{\mathbf{J}} = \frac{1}{V} \left(\int_{\Omega} \mathbf{j}(\mathbf{x})d\Omega + \int_{\Gamma} \mathbf{j}^{s}(\mathbf{x})d\Gamma \right), \tag{5}$$

where $\mathbf{j}^{s}$ is a surface electric flux (see Section 3). In the so-called FE^2 method, the constitutive law $\overline{\mathbf{J}}$ - $\overline{\mathbf{E}}$ is unknown, but can be numerically obtained by solving a nonlinear problem over the RVE, detailed as follows (see Figure 1):

Given $\overline{\mathbf{E}}$:

1. Prescribe boundary conditions (4) on $\partial\Omega$.
2. Use a numerical method such as FEM with an iterative solver such as the Newton method to solve nonlinear Problems (1), (2), and (4) (see details in the following).
3. Compute the spatial average of the electric flux over the RVE to obtain $\overline{\mathbf{J}}$.

In what follows, a detailed numerical implementation of a FE^2 problem in a context of nonlinear electric conduction is presented to better understand where Problems (1), (2), and (4) must be solved within finite-element calculation at the macroscopic scale. The macroscopic problem at the macroscale is given by

$$\nabla \cdot \overline{\mathbf{J}} = 0 \quad \text{in } \overline{\Omega}, \tag{6}$$

with boundary conditions:

$$\overline{\phi} = \overline{\phi}^* \text{ on } \partial\overline{\Omega}_\phi, \ \ \overline{\mathbf{J}} \cdot \mathbf{n} = \overline{J}_n^* \text{ on } \partial\overline{\Omega}_J, \tag{7}$$

where and $\overline{\Omega}_\phi$ and $\partial\overline{\Omega}_J$ denote the Dirichlet and Neumann complementary boundaries, respectively.

In what follows, we assume $\overline{J}_n^* = 0$. Then, the weak form corresponding to (6) is given by:

$$\int_{\overline{\Omega}} \overline{\mathbf{J}}(\overline{\phi}) \cdot \nabla(\delta\overline{\phi}) d\Omega = \overline{R}(\overline{\phi}) = 0. \tag{8}$$

Problem (8) is nonlinear. Then, a Newton method is employed to solve it. A first-order Taylor expansion of $\overline{R}(\overline{\phi})$ gives

$$\overline{R}(\overline{\phi}^k + \Delta\overline{\phi}) \simeq \overline{R}(\overline{\phi}^k) + D_{\Delta\overline{\phi}}\overline{R}(\overline{\phi}^k), \tag{9}$$

where $\overline{\phi}^k$ is a solution provided at a previous iteration, and $D_{\Delta\overline{\phi}}\overline{R}(\overline{\phi})$ is the Gateaux derivative, expressed by

$$D_{\Delta\overline{\phi}}\overline{R}(\overline{\phi}^k) = \left[\frac{d}{d\alpha}\left(\overline{R}(\overline{\phi} + \alpha\Delta\overline{\phi}) \right) \right]_{\alpha=0}. \tag{10}$$

The corresponding linearized problem is given by

$$D_{\Delta\overline{\phi}}\overline{R}(\overline{\phi}^k) = -\overline{R}(\overline{\phi}^k), \tag{11}$$

with

$$D_{\Delta\overline{\phi}}\overline{R}(\overline{\phi}^k) = -\int_{\Omega} \overline{\overline{\mathbf{k}}}(\phi^k)\nabla(\Delta\overline{\phi}) \cdot \nabla(\delta\phi) d\Omega. \tag{12}$$

More details can be found in [48]. Classical FEM discretizing of (11) leads to linear system

$$\overline{\mathbf{K}}_T(\overline{\phi}^k)\Delta\overline{\phi} = -\mathbf{R}(\overline{\phi}^k). \tag{13}$$

Then, the macroelectric potential is updated according to

$$\overline{\phi}^{k+1} = \overline{\phi}^k + \Delta\overline{\phi} \tag{14}$$

and (13) is solved until a convergence criterion is reached. In FE^2, the main source of computational cost is the numerical evaluation of $\overline{\mathbf{J}}$ and $\overline{\mathbf{k}}$, obtained by solving nonlinear RVE Problems (1), (2), and (4) at each Gaussian point. To address this issue, we introduce a data-driven approach where the estimation of $\overline{\mathbf{J}}$ is provided by a neural-network-based surrogate model. Tangent matrix $\overline{\mathbf{k}}$ can be computed by a perturbation approach as

$$\left(\overline{k}_T \right)_{ij}(\overline{\mathbf{E}}) = \frac{\partial \overline{J}}{\partial \overline{E}} \simeq \frac{\overline{J}_i(\overline{\mathbf{E}} + \delta\overline{\mathbf{E}}^{(j)}) - \overline{J}_i(\overline{\mathbf{E}})}{\alpha} \tag{15}$$

with

$$\delta\overline{\mathbf{E}}^{(j)} = \alpha\mathbf{e}_j, \tag{16}$$

where $\mathbf{e}_j$ is a unitary vector basis, and $\alpha << 1$ a perturbation parameter.

Then, to compute macro-FEM nonlinear calculations, relationship $\overline{\mathbf{J}} = \overline{\mathcal{F}}^{nl}(\overline{\mathbf{E}})$ is missing. In [30], we proposed a surrogate model to construct such a relationship using neural networks. In the present paper, this idea is extended to random microstructures, as detailed in the next section.

3. Micro-Non-Linear Conduction Model for Graphene-Reinforced Composites

In this section, the nonlinear conduction model in graphene-reinforced polymer composites is defined. The nonlinear RVE problem is described as follows. The microstructure was assumed to be characterized by an RVE defined in a domain $\Omega \subset \mathbb{R}^3$ that contained N randomly distributed planar multilayer graphene sheets (see Figure 2). The graphene sheets were assumed to be aligned along the $x - y$ plane. We chose this configuration for two reasons: (i) when samples made of graphene-reinforced polymer are obtained via injection molding, the graphene sheets can be aligned in the direction of the polymer flow [49]. Then, this configuration is representative of samples manufactured by the injection-molding process. Second, such an orientation induces strong anisotropy of the effective nonlinear conductive behavior of the material. The potential of the present approach to deal with such a challenging problem is then illustrated.

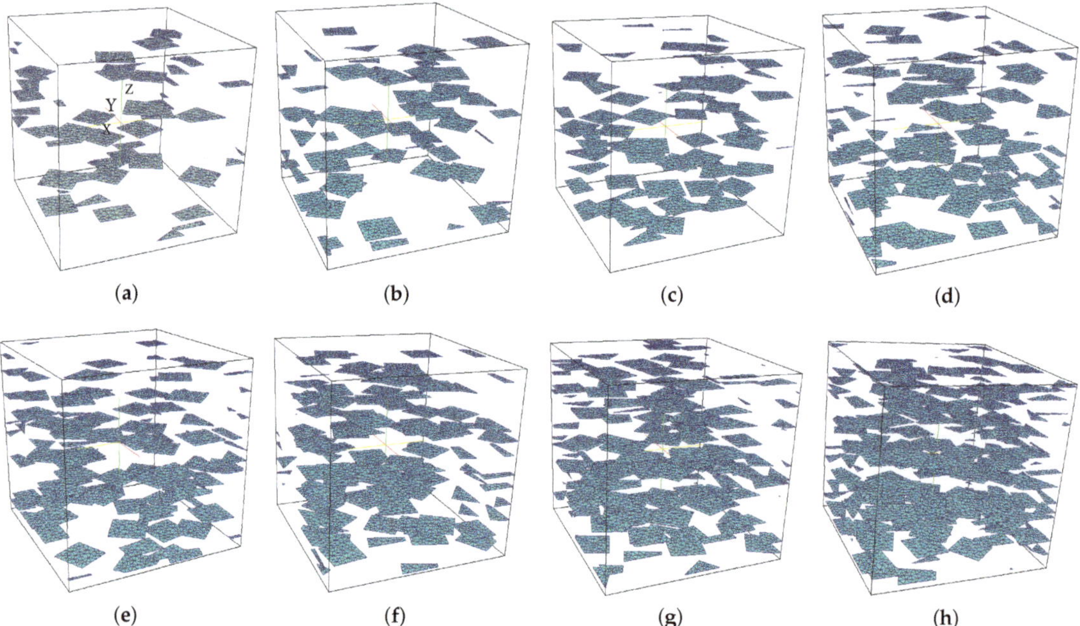

Figure 2. Realizations of microscopic RVE with various graphene volume fractions: (**a**) 0.53 vol%, (**b**) 0.66 vol%, (**c**) 0.79 vol%, (**d**) 0.92 vol%, (**e**) 1.05 vol%, (**f**) 1.19 vol%, (**g**) 1.32 vol%, (**h**) 1.58 vol% [30].

To avoid meshing their thickness [48], the graphene sheets were modeled as highly conducting imperfect surfaces here [50]. The graphene surfaces with zero thickness are collectively denoted by Γ. The nonlinear behavior is related to the electric tunnelling effect here, which may be an explanation for the observed nonlinear behavior and low percolation thresholds in the nanocomposites (see [30,48]).

The energy of the system is defined by

$$W = \int_\Omega \omega^b(\mathbf{x})d\Omega + \int_\Gamma \omega^s(\mathbf{x})d\Gamma, \tag{17}$$

where density functions ω^b and ω^s are the bulk and surface density functions, respectively, expressed by

$$\omega^b(\mathbf{x}) = \frac{1}{2}\mathbf{j}(\mathbf{x}) \cdot \mathbf{E}(\mathbf{x}), \quad \omega^s(\mathbf{x}) = \frac{1}{2}\mathbf{j}^s(\mathbf{x}) \cdot \mathbf{E}^s(\mathbf{x}). \tag{18}$$

In (18), $\mathbf{E}^s(\mathbf{x})$ and $\mathbf{j}^s(\mathbf{x})$ are the surface electric field and surface current density with respect to the graphene sheets, where $\mathbf{E}^s = \mathbf{P}\mathbf{E} = -\mathbf{P}\nabla\phi$, where $\mathbf{P} = \mathbf{I} - \mathbf{n}\otimes\mathbf{n}$ is a projector operator characterizing the projection of a vector along the tangent plane to Γ at a point $\mathbf{x} \in \Gamma$, and $\mathbf{n}$ is the unit normal vector to Γ.

The nonlinear electric-conduction law including the tunneling effect in the polymer matrix is given by

$$\mathbf{j} = \begin{cases} \mathbf{k}_0^p \mathbf{E} & \text{if } d(\mathbf{x}) \le d_{cut}, \\ \mathcal{G}(\mathbf{E},d)\frac{\mathbf{E}}{|\mathbf{E}|} & \text{if } d(\mathbf{x}) > d_{cut}, \end{cases} \tag{19}$$

where d_{cut} is a cut-off parameter, and $\mathbf{k}_0^p$ is the electric-conductivity tensor of the polymer matrix without tunneling effects. The distance function between graphene sheets $d(\mathbf{x})$ is defined here as the sum of the two smallest distances between a point $\mathbf{x}$ in the polymer matrix and the two nearest-neighbor graphene sheets (see more details in [48]). Nonlinear tunneling current $\mathcal{G}$ was proposed by Simmons [51] as

$$\mathcal{G}(E,d) = \frac{2.2e^3 E^2}{8\pi h_p \Phi_0} \exp\left(-\frac{8\pi}{2.96h_p eE}(2m)^{\frac{1}{2}}\Phi_0^{\frac{3}{2}}\right)$$

$$+ \left[3 \cdot \frac{(2m\Phi_0)^{\frac{1}{2}}}{2d}\right](e/h_p)^2 E d \exp\left[-\left(\frac{4\pi d}{h_p}\right)(2m\Phi_0)^{\frac{1}{2}}\right]. \tag{20}$$

Above, Φ_0 and d denote barrier height and barrier width, respectively, e and m are the charge and the effective mass of electron, respectively, and h_p is the Planck constant. Surface current density $\mathbf{j}^s$ of graphene surface Γ is related to surface electric field $\mathbf{E}^s$ [50] through

$$\mathbf{j}^s(\mathbf{x}) = \mathbf{k}^s \mathbf{E}^s, \tag{21}$$

where $\mathbf{k}^s$ denotes the the surface conductivity of the graphene. This tensor can be related to the conductivity of the volume (multilayer) graphene as:

$$\mathbf{k}^s = h\mathbf{S}^*, \quad \mathbf{S}^* = \mathbf{k}^g - \frac{(\mathbf{k}^g\mathbf{n})\otimes(\mathbf{k}^g\mathbf{n})}{\mathbf{k}^g : (\mathbf{n}\otimes\mathbf{n})}. \tag{22}$$

where h is the thickness of the graphene sheet.

Considering the constitutive equations above, and minimizing the electric power (17) with respect to the electric potential field, the weak form is obtained as

$$\int_\Omega \mathbf{j}(\phi) \cdot \nabla(\delta\phi)d\Omega - \int_\Gamma \mathbf{P}\nabla\phi \cdot \mathbf{k}^s \mathbf{P}\nabla(\delta\phi)d\Gamma = 0, \tag{23}$$

where $\phi \in H^1(\Omega)$, ϕ satisfying the Dirichlet boundary conditions over $\partial\Omega$ and $\delta\phi \in H^1(\Omega)$, $\delta\phi = 0$ over $\partial\Omega$. The RVE is subjected to boundary conditions (4). The weak form (23) can be solved by the finite-element method.

4. Stochastic Nonlinear Machine-Learning Model

The objective of the present work was to construct a surrogate model relating macroscopic electric field $\bar{\mathbf{E}}$ and volume fraction of graphene inclusions f to nonlinear macroscopic electric flux response $\bar{\mathbf{J}}$ (see Figure 3). At the microscale, the microstructure was randomly distributed (see Figure 2). Here, due to the scale-separation assumption, it was expected that, despite the random nature of the microstructure, deterministic effective properties at the microscopic scale with respect to the microstructure would be obtained for a given volume fraction.

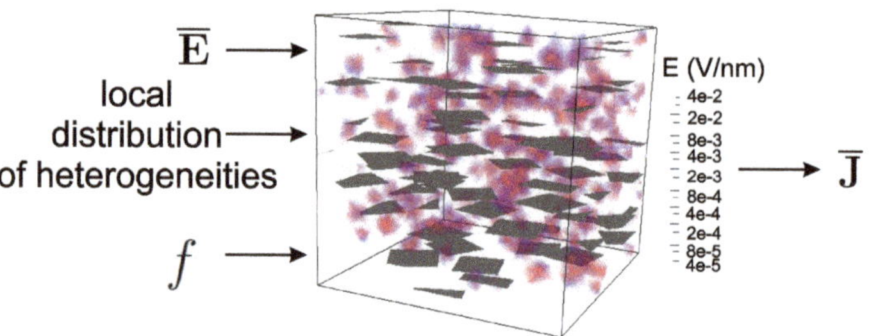

Figure 3. Local model: effective flux $\bar{\mathbf{J}}$ depends nonlinearly on the macroscopic electric field $\bar{\mathbf{E}}$, volume fraction, and local random distribution of phases.

At the macroscale, uncertainties where then only related to nonhomogeneous distributions of volume fractions. Then, it was assumed that, at the macroscale, the volume fraction was the only stochastic parameter.

4.1. Data Generation

We first define a set of K electric-field vectors, $\bar{\mathbf{E}}^k = \{\bar{E}_1, \bar{E}_2, \bar{E}_3\}^k$, $(k = 1, 2, \ldots, K)$. The values of $\bar{\mathbf{E}}^k$ were generated using Latin hypercube sampling (LHS) [52]. Then, we define a collection of microstructures as follows. A set of P volume fractions are defined, f^α, $\alpha = 1, 2, \ldots, P$. For each volume fraction f^α, N random microstructures satisfying volume fraction f^α were generated and are denoted by Ω^i_α, $i = 1, 2, \ldots, N$.

Then, for each macroelectric field vector $\{\bar{\mathbf{E}}\}^k$, each volume fraction f^α and each realization of microstructure Ω^i_α, nonlinear problem (23) is solved by finite elements to obtain macroelectric displacement vector $\{\bar{\mathbf{J}}\}^k_{\alpha,i}$.

As discussed above, the scale-separation assumption allows for removing the stochastic nature of the microstructures at the macroscale. However, due to the RVE size and the random distribution of the inclusions, the outcome intensity of a given electric field is also stochastic. To this purpose, homogenization is performed using stochastic averaging, i.e., for each macroelectric field vector $\bar{\mathbf{E}}^k$ and each volume fraction f^α, we compute the average over N microstructures realizations to obtain $\bar{\mathbf{J}}^k_\alpha = \frac{1}{N} \sum_{i=1}^{N} \left(\bar{\mathbf{J}}^k_{\alpha,i} \right)$.

4.2. Construction of Neural-Network Surrogate Model

An issue in NN models is that a large set of data may be required to obtain good accuracy, especially for a large number of input parameters [28]. To overcome this limitation, we propose here a hybrid NN/interpolation surrogate model as follows.

First, for each volume fraction f^α, $\alpha = 1, 2, \ldots, P$, used in the training dataset, we define a separate NN, denoted by $\mathcal{N}^\alpha$, in order to construct the following relationship:

$$\bar{\mathbf{J}}_\alpha(\bar{\mathbf{E}}) = \mathcal{N}^\alpha(\bar{\mathbf{E}}). \tag{24}$$

Then, given $\bar{\mathbf{E}}$ and for an arbitrary volume fraction $f \in \left[f^1, f^P \right]$ a Lagrangian interpolation scheme is used to compute $\bar{\mathbf{J}}(\bar{\mathbf{E}}, f)$ as

$$\bar{\mathbf{J}}(\bar{\mathbf{E}}, f) = \sum_{j \in N_k(f)} l_j(f) \bar{\mathbf{J}}_j(\bar{\mathbf{E}}), \tag{25}$$

where $N_k(f)$ is the set of indices that includes only k out of P NNs, corresponding to the k volume fractions of f nearest to those in training dataset $\{f^1, f^2, \ldots, f^P\}$. In Equation (25), $l_j(f)$ are the Lagrangian basis polynomials. Here, $k = 3$ was employed where, as a result, polynomials $l_j(f)$ were second-order.

With this approach, a notion of locality is introduced in the interpolation scheme that leads to better overall predictions, especially in areas where relationship $(\bar{\mathbf{E}}, f)$ –

$\bar{J}(\bar{E}, f)$ exhibits strong nonlinearity. A schematic of the surrogate-model construction is summarized in Figure 4.

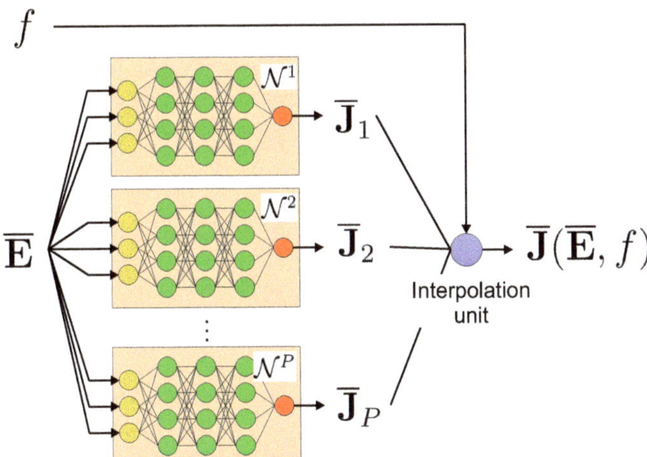

Figure 4. Hybrid neural-network/interpolation surrogate model to describe macroscopic nonlinear behavior.

5. Nonlinear Stochastic Macroscale Calculations

Here, stochastic macroscopic structural problem is described. At the macroscale, it was assumed that there existed uncertainty in the local volume fraction f of the graphene sheets in the general case described by a 3D homogeneous Gaussian stochastic field in the x, y, and z axes. In particular, if $x \equiv (x, y, z)$, then $f(x)$ is considered to be of the form:

$$f(x, \theta) = \mu + \sigma f_0(x, \theta). \tag{26}$$

In the above equation, μ and σ are the random field mean value and standard deviation, respectively, θ denotes the random outcome, and $f_0(x, \theta)$ is a zero-mean unit variance Gaussian field with correlation structure R_{f_0} given by

$$R_{f_0}(x_1, x_2) = \exp\left[-\left(\frac{|x_1 - x_2|}{\hat{a}_x} + \frac{|y_1 - y_2|}{\hat{a}_y} + \frac{|z_1 - z_2|}{\hat{a}_z}\right)\right] \tag{27}$$

where $\hat{a}_x, \hat{a}_y$ and $\hat{a}_z$ are correlation length parameters along the x, y, and z axes, respectively.

Next, an approximation of field f_0 can be obtained using the Karhunen–Loeve series expansion [53]. Specifically, let λ_n, ϕ_n denote the eigenvalues and eigenfunctions that satisfy the eigenvalue problem $\int R_{f_0}(x_1, x_2)\phi_n(x_2)dx_2 = \lambda_n\phi_n(x_1)$, $\forall n = 1, \ldots.$ This is a Fredholm integral equation and it is typically solved using the finite-element method [53]. Then, f_0 can be written as

$$f_0(x, \theta) = \sum_{n=1}^{\infty} \sqrt{\lambda_n} z_n(\theta)\phi_n(x) \tag{28}$$

with $\{z_n\}_{n=1}^{\infty}$ being a series of uncorrelated Gaussian random variables with zero mean and unit variance. In practice, the above series expansion is truncated after M_{KL} terms, giving the following approximation of f_0.

$$f_0(x, \theta) \approx \sum_{n=1}^{M_{KL}} \sqrt{\lambda_n} z_n(\theta)\phi_n(x), \tag{29}$$

which yields, by virtue of Equation (26),

$$f(\pmb{x},\theta) \approx \mu + \sigma \sum_{n=1}^{M_{KL}} \sqrt{\lambda_n} z_n(\theta) \phi_n(\pmb{x}). \tag{30}$$

Equation (30) allows for us to generate realizations of the field $f(\pmb{x},\theta)$ by generating M_{KL}-tuples $(z_1,\ldots,z_{KL})$ from their distribution. Subsequently, if we consider macrostructure $\overline{\Omega}$ defined in Section 2 and the associated finite-element mesh, at each Gaussian point of element $\overline{\Omega}^e$, $e = 1,2,\ldots,N_e$ with coordinate $\pmb{x}$, a random value of volume fraction $f(\pmb{x})$ can be assigned using (30).

During the Newtonian procedure to solve the structural problem, for f and $\overline{\mathbf{E}}$ given at each Gaussian point of the macromesh structure, the corresponding value of $\overline{\mathbf{J}}$ is provided by the surrogate model (25) (see Figure 4). For one realization of the volume-fraction distribution generated by Equation (30), the cost of one two-scale nonlinear structural problem is drastically reduced with the present NN surrogate model, allowing for performing a large number of macrocalculations at a low cost to conduct statistics on quantities of interest in a structure.

Lastly, Monte Carlo simulations were performed on the macroscale problem by evaluating R realizations of macrostructures. For each realization $r = 1,2,\ldots,R$, the volume fraction was randomly generated in each Gaussian point by using Equation (30); in total, R nonlinear multiscale problems were solved using the above-described procedure. Lastly, statistics can be computed on quantities of interest using the R nonlinear FEM solutions. The overall procedure is summarized in Figure 5.

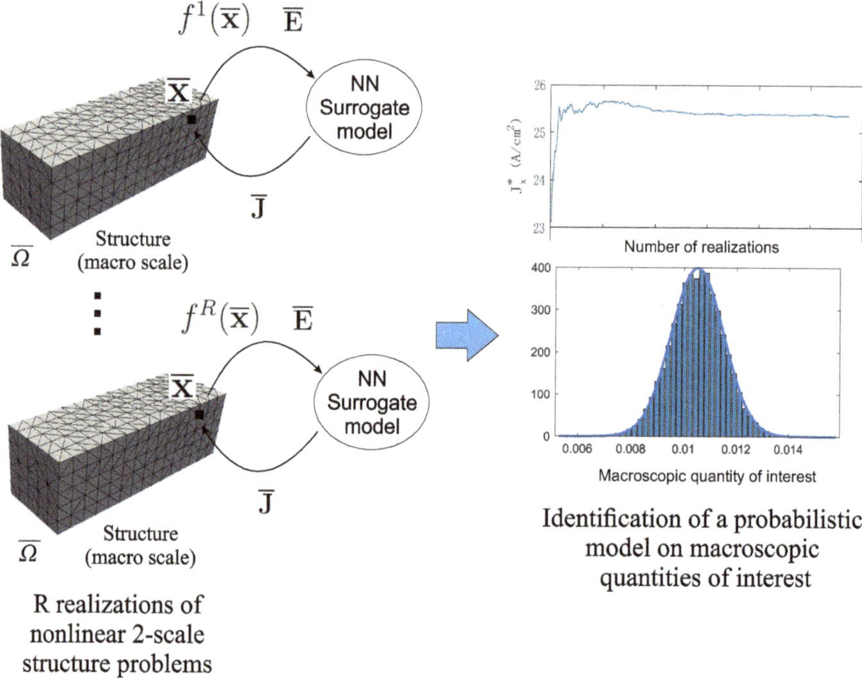

Figure 5. Stochastic nonlinear 2-scale procedure.

6. Numerical Examples

6.1. Data Collection

The data were obtained by performing preliminary calculations on the RVE described in Section 4.1. Eight different volume fractions were considered: $f^1 = 0.53\%$, , $f^2 = 0.66\%$, $f^3 = 0.79\%$, $f^4 = 0.92\%$, $f^5 = 1.05\%$, $f^6 = 1.19\%$, $f^7 = 1.32\%$, and $f^8 = 1.58\%$ (see Figure 2). For each volume fraction, 15 realizations of random microstructures were generated except for the higher volume fraction, for which only 9 realizations were conducted. For each volume fraction and for each realization, 500 realizations of macroscopic electric field $\bar{E}$ were generated using Latin hypercube sampling. For each case, corresponding electric flux $\bar{J}$ was numerically computed by solving nonlinear Problem (23) on the RVE. The total number of solved nonlinear problems was then 57,000. All these calculations could be performed in parallel.

6.2. Validation of Hybrid NN–Interpolation Surrogate Model

The accuracy of the proposed hybrid NN–interpolation surrogate model was first validated by comparing its response with full-field simulations on microstructures for different volume fractions. Regarding the characteristics of the trained neural networks, in all cases, one-hidden-layer architectures were considered with the optimal number of neurons varying for each case, as shown in Table 1. Moreover, the hyperbolic tangent function was employed as the activation function, and Levenberg–Marquardt as the optimizer in all NNs.

Table 1. Characteristics of neural networks.

Case (Vol%)	Number of Neurons	MSE (Validation Set)
0.53	16	1.167×10^{-18}
0.66	23	6.785×10^{-18}
0.79	7	4.006×10^{-12}
0.92	18	8.941×10^{-7}
1.05	77	1.579×10^{-5}
1.19	59	3.465×10^{-4}
1.32	36	1.243×10^{-1}
1.58	74	4.790×10^{-2}

The plotted curves were obtained as the average over the different realizations of the microstructure. Results are provided in Figure 6. For low volume fractions, the response was linear, while for larger volume fractions, the response was strongly nonlinear. In all cases, the surrogate model could accurately reproduce the effective nonlinear response of the material.

A validation of the interpolation procedure described in Section 4.2 is provided in Figure 7, where discrete data obtained by nonlinear FEM calculations on the RVEs are compared to the corresponding model predictions, computed using Equation (25) under various $\bar{E}_x$ scenarios, with $\bar{E}_y = \bar{E}_z = 0$. The discrete data points obtained by FEM are denoted by marks, while the continuous interpolation with respect to the volume fraction is denoted by solid lines, which confirmed the good accuracy of this scheme.

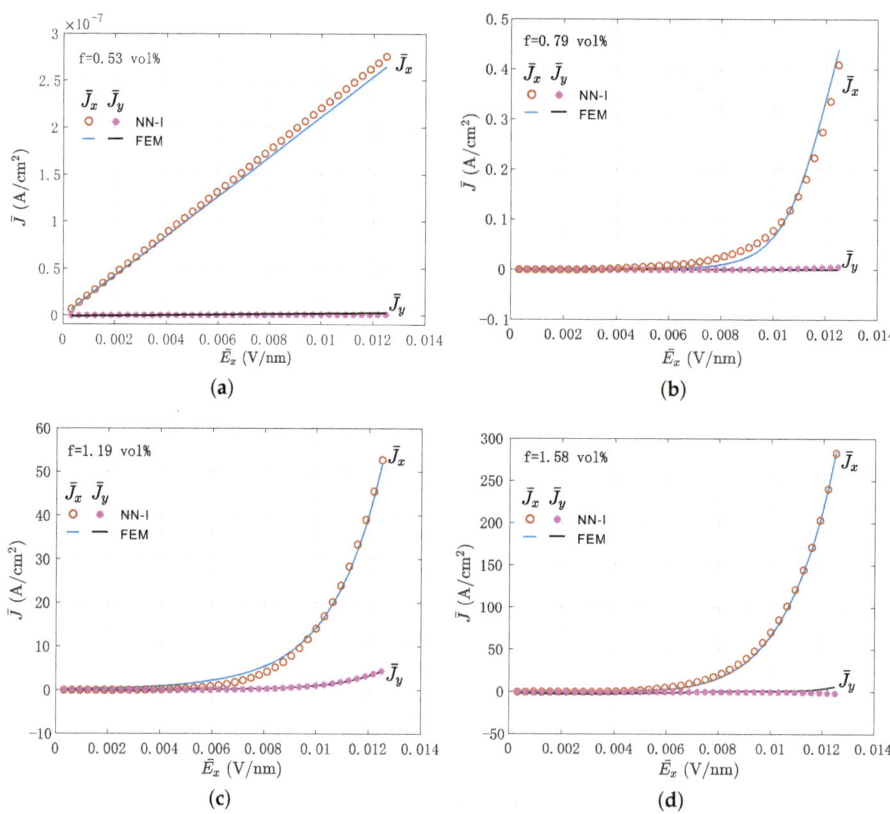

Figure 6. Comparisons between direct simulations obtained by nonlinear FEM calculations on the RVE and the neural-network–interpolation surrogate model: values of $\bar{J}_x$ as a function of a unidirectional effective electric field $\bar{E}_x$, $\bar{E}_y = \bar{E}_z = 0$; (**a**) 0.53 vol%, (**b**) 0.79 vol%, (**c**) 1.19 vol%, (**d**) 1.58 vol%.

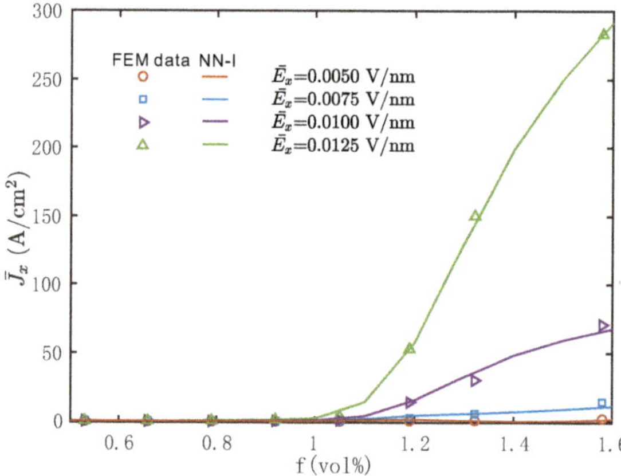

Figure 7. Comparisons between direct simulations obtained by nonlinear FEM calculations on the RVE and NN–I model under various E_x ranging from 0.0050 to 0.0125 V/nm: values of J_x as a function of the CNT volume fraction, $E_y = E_z = 0$.

6.3. Stochastic 2-Scale Nonlinear Structure Analysis

In this example, macroscopic stochastic nonlinear computations were performed using the procedure described in Section 5. In particular, 9 different Gaussian fields of volume fractions are investigated at the macroscale, where the studied macrostructure, described in Figure 8, was a plate with a central hole. The plate was subjected to potential boundary conditions such as $\Phi = \Phi_1$ on $x = 0$ and $\Phi = \Phi_2$ on $x = L$. A 3D mesh of 1934 elements is used to discretize the domain.

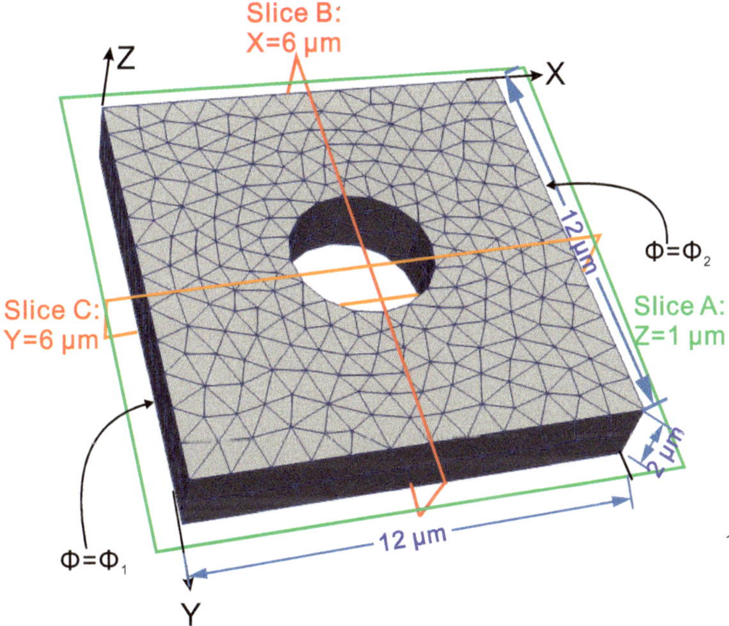

Figure 8. Structural problem: geometry, boundary conditions, and mesh.

Due to the low thickness of the structure, we assumed that the volume fraction did not vary in the z coordinate direction. Next, in order to define the aforementioned Gaussian fields, three different settings were first initialized: for Setting A, we set $\mu^A = 0.9\%$ and $\sigma^A = 0.11\,\mu^A$; for Setting B, we set $\mu^B = 1.05\%$ and $\sigma^B = 0.19\,\mu^B$; lastly, for Setting C, $\mu^C = 1.05\%$ and $\sigma^C = 0.38\,\mu^C$. Then, for each aforementioned setting, we considered $\hat{a}_z = \hat{a}_y = \hat{a}$ and assign three different values to $\hat{a}$, namely, $\hat{a}_1 = 6\,\mu m$, $\hat{a}_2 = 12\,\mu m$ and $\hat{a}_3 = 24\,\mu m$. A sample for each of these fields is illustrated in Figure 9. This figure indicates that an increase in the field standard deviation led to larger variations of volume fraction f along the spatial domain. Moreover, a small correlation-length parameter, such as $\hat{a} = 6\,\mu m$, produced more "wavy" realizations, while for larger values ($\hat{a} = 12$ and $24\,\mu m$) the realizations became smoother.

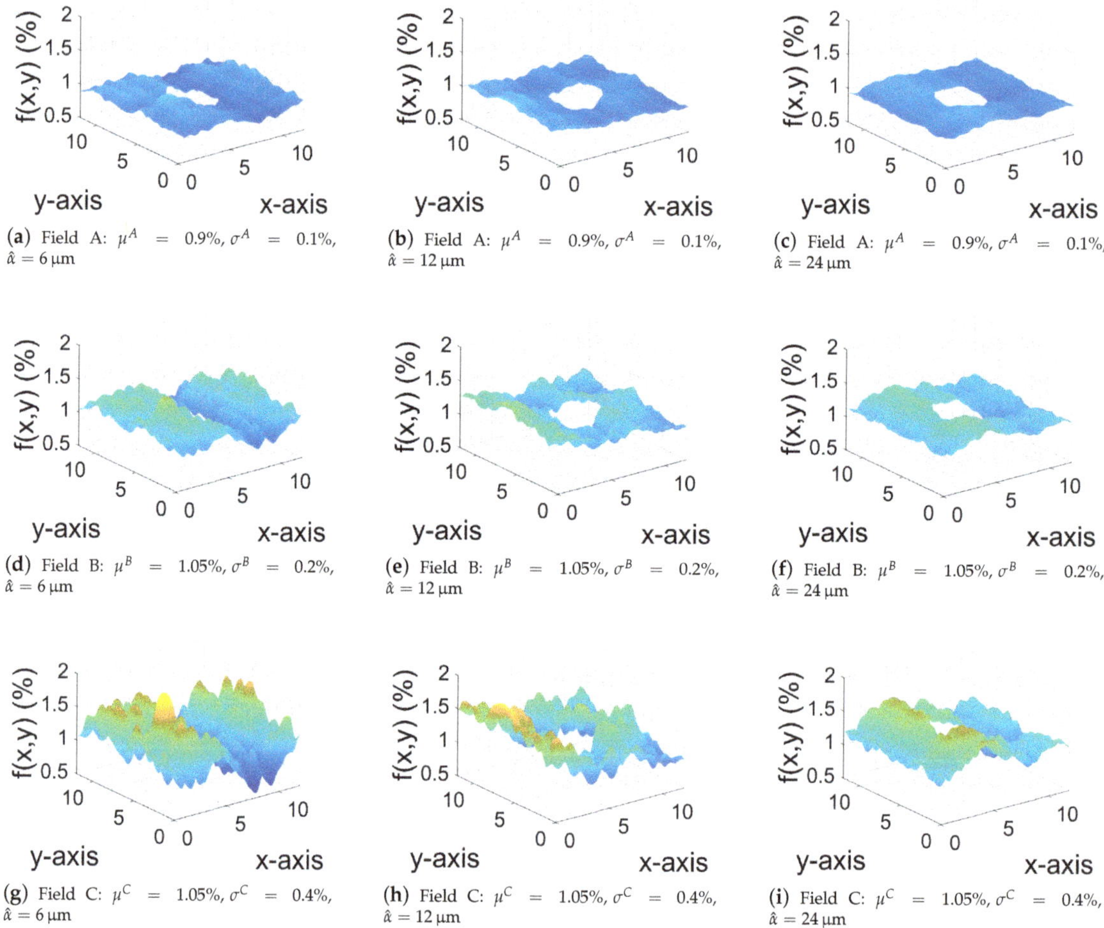

Figure 9. Sample realizations of three Gaussian fields A, B, and C for different correlation lengths $\hat{a} = 6, 12, 24\,\mu m$.

For each of the 3 Gaussian distributions A, B, and C, we analyzed the 3 correlation lengths $\hat{a}_1$, $\hat{a}_2$ and $\hat{a}_3$. For each case, we conducted 100 realizations. Then, in total, we conducted 900 FE^2-NN simulations using the procedure described in Section 5. For each one, a stochastic distribution of volume fraction was generated in the elements using (30). The macroscopic quantity of interest is defined here as the average macroscopic flux in the domain $\overline{\Omega}$ as

$$\mathbf{J}^* = \frac{1}{\overline{V}} \int_{\overline{\Omega}} \bar{\mathbf{J}} d\Omega, \tag{31}$$

where $\overline{V}$ is the volume of $\overline{\Omega}$. The convergence of the components of $\mathbf{J}^*$ is depicted in Figure 10. In all cases, statistical convergence could be achieved. For the lowest average values f and standard deviation σ of the volume fractions (Cases 1–3 in Figure 10), correlation length $\hat{a}$ did not have significant influence on the convergence rate. However, for larger values of f and σ, convergence could be much slower (e.g., Case 9 in Figure 10), where around 50 realizations are necessary to achieve convergence. This clearly illustrates the interest of the proposed surrogate-based multiscale method, where each realization is performed at the cost of a classical FEM simulation. In contrast, using standard FE^2 would not allow performing this kind of statistical analysis with available computer resources.

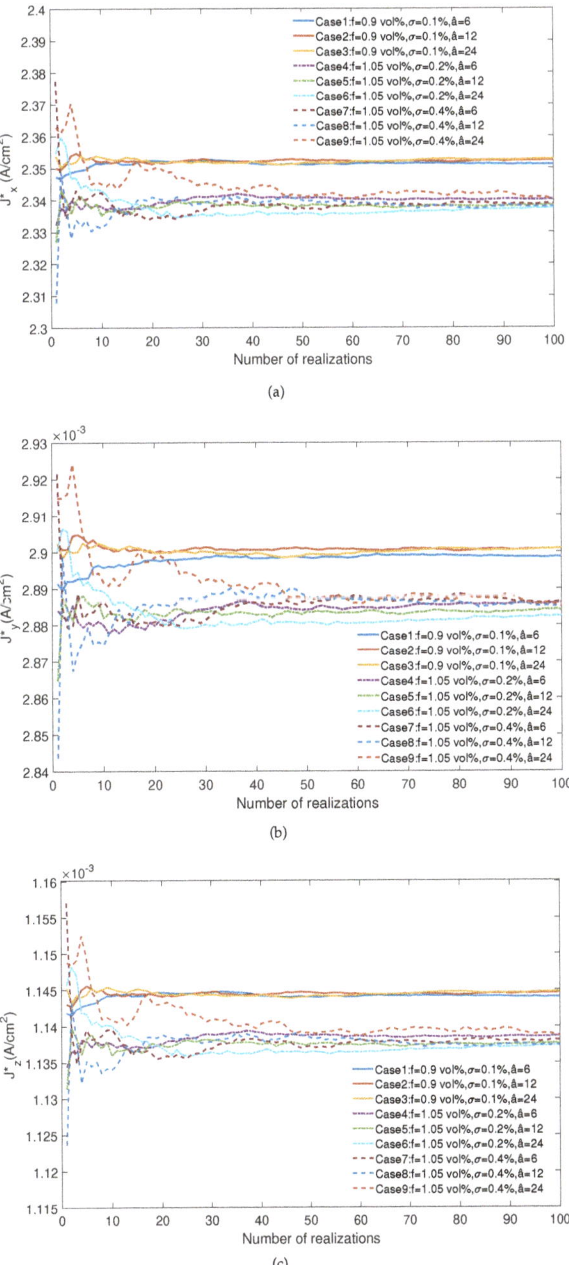

Figure 10. Averaged current-density components as a function of the number of realizations under various distributions of CNT volume fraction and different correlation lengths. (**a**) J_x^*; (**b**) J_y^*; (**c**) J_z^*.

Average distributions of local current densities over 100 realization are plotted in Figure 11 corresponding to distribution A and correlation length $\hat{a}$=6 μm. Clear anisotropy of the effective behavior induced by the aligned graphene sheets along the $x - y$ plane can be appreciated. Comparing Figure 11a,b, we can see a clear difference in the magnitude of the $\bar{J}_x$ and $\bar{J}_z$ values, indicating that the effective conductivity in the z direction was much lower than that in the $x - y$ plane. The present method could capture such anisotropic behavior in a nonlinear stochastic context.

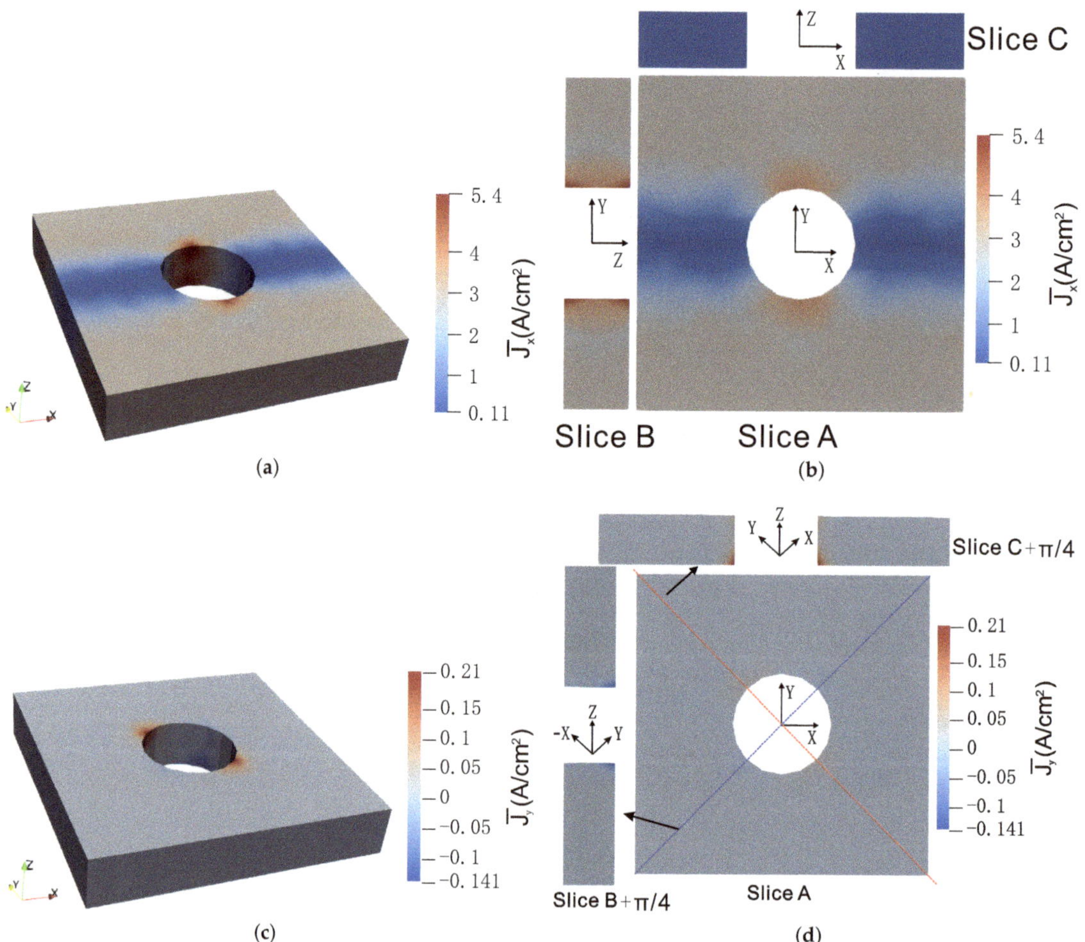

Figure 11. Averaged current density $\bar{J}_x$ and $\bar{J}_y$ over 100 realizations calculated by the NN–I model for the composite structure for potential difference $\Phi_2 - \Phi_1 = 144$ V. The CNT volume fraction obeys distribution A with $\mu^A = 0.9$ vol%, $\sigma^A = 0.1\%$, and correlation length $\hat{a}$=6 μm: (**a**) $\bar{J}_x$-component: 3D view; (**b**) $\bar{J}_x$-component: plots along different planes; (**c**) $\bar{J}_y$-component: 3D view; (**d**) $\bar{J}_y$-component: plots along different planes.

The evolution of the quantity of interest J_x^* was plotted with respect to the difference of the potential applied over macrostructure $\Phi_2 - \Phi_1$ in Figure 12. Various distributions of CNT volume fractions and different correlation lengths were taken into account for comparison. For each case, 100 realizations were computed, from which we obtained the average and deviation of J_x^*. For instance, in Figure 12a, correlation length $\hat{a} = 6$ μm is for all three different CNT volume-fraction distributions. The averaged value of J_x^* was independent on standard deviation σ of the Gaussian distribution, whereas the deviation of J_x^* increased slightly with increasing σ. The same phenomenon could also be observed in Figure 12b,c. Furthermore, by comparing Figure 12a–c, the increase in correlation length led to a tiny increase in the deviation of J_x^*, but had no effect on its averaged value.

Lastly, in Figure 13, distributions of target values J_x^*, J_y^* and J_z^* are plotted for selected cases of the probabilistic models describing the distribution of the volume fraction in the macroscale. In Figure 14, the associated empirical cumulative distribution functions (ECDFs) are provided. These functions were identified from the histograms in Figure 13. These allow for a direct quantitative reading of key values of interest (minimum, maximum, mean, percentiles, etc.) regarding the macroscopic quantities. ECDFs also have the property of converging to the true CDF of the stochastic quantities of interest as the number of samples is increased [54]. Typically, an accurate estimate of a CDF would require a very large number of samples ($>10^5$); however, performing these many evaluations of nonlinear multiscale models would be computationally prohibitive. In this regard, the use of the proposed surrogate is the only viable approach to obtain reliable approximations of the CDFs under investigation. This demonstrates the potential of the present approach in constructing probabilistic models for macroquantities of interest in nonlinear multiscale models of random materials.

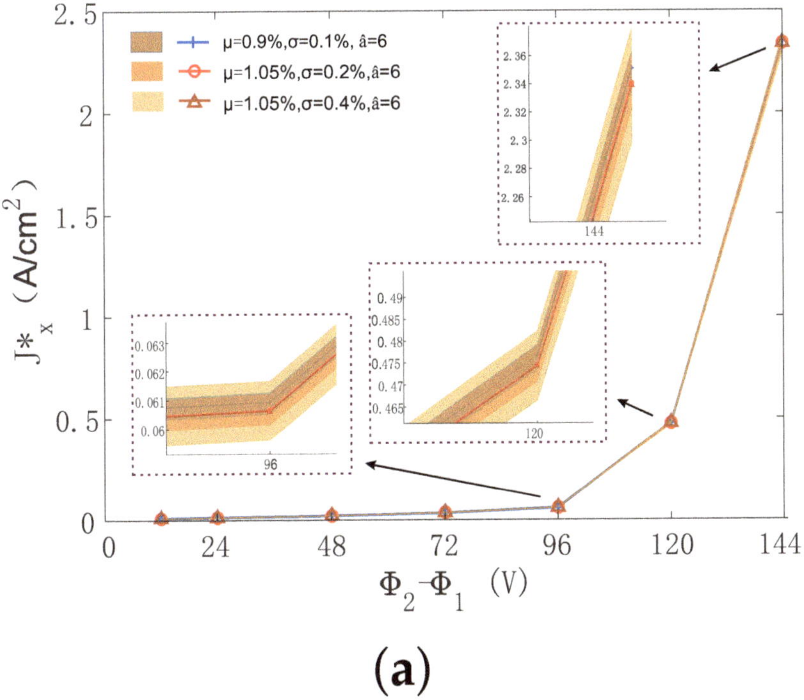

(a)

Figure 12. *Cont.*

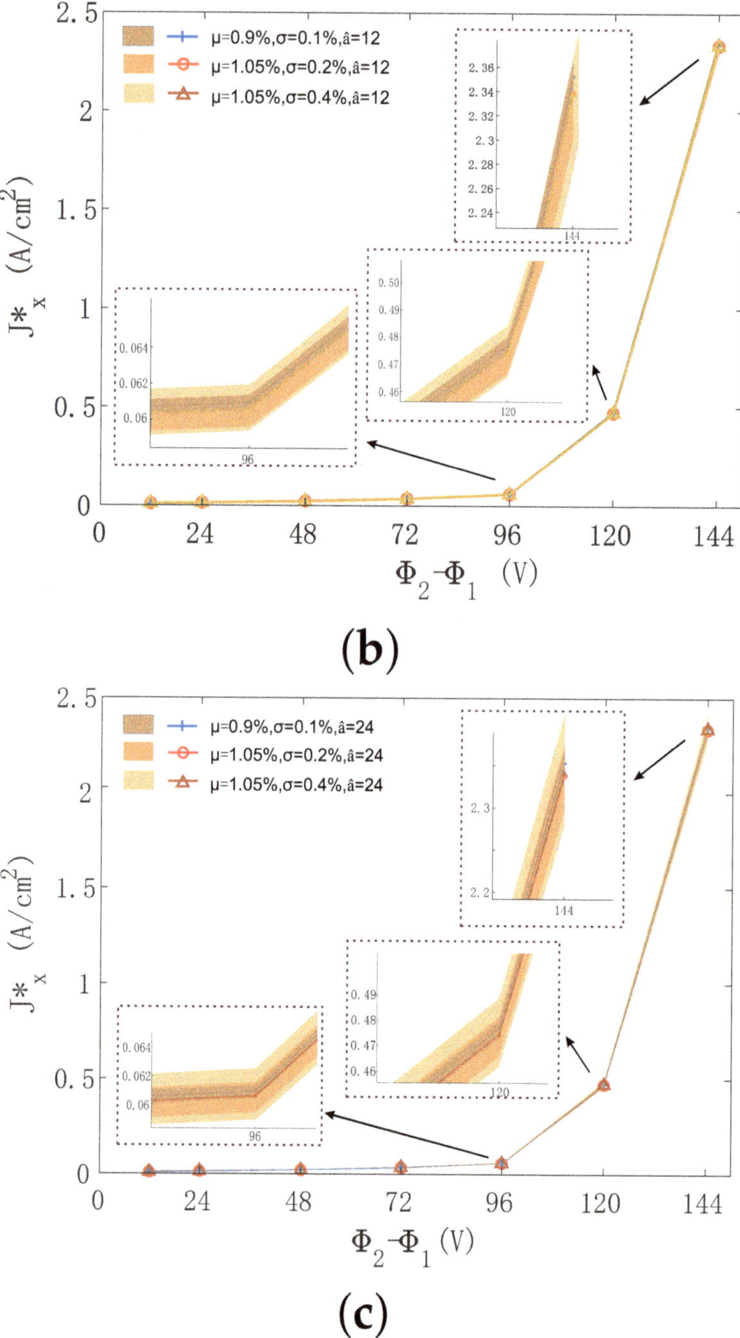

(b)

(c)

Figure 12. Averaged current density J_x^* and corresponding deviation as a function of potential difference $\Phi_2 - \Phi_1$ for various distributions of CNT volume fraction under different correlation lengths $\hat{a}$. (**a**) $\hat{a}_1 = 6$ μm; (**b**) $\hat{a}_2 = 12$ μm; (**c**) $\hat{a}_3 = 24$ μm. Color zones indicate ranges of values.

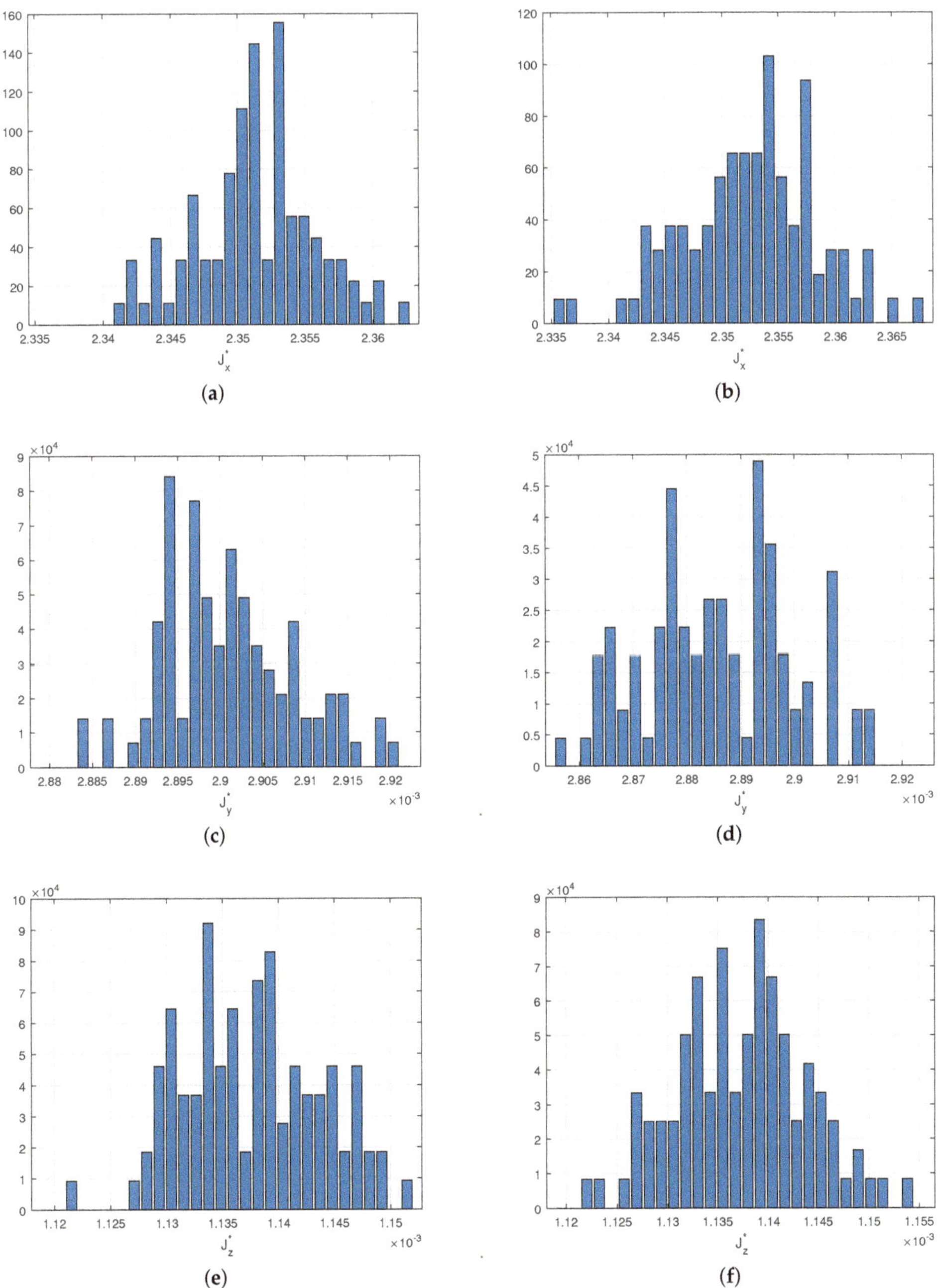

Figure 13. Histograms associated to probabilistic models describing distribution of $\overline{\mathbf{J}}^*$ components at the macroscale. Values of $\overline{\mathbf{J}}^*$ are reported for a fixed value of macroboundary condition $\Phi_2 - \Phi_1 = 144V$; (**a**) J_x^*, $\mu = 0.9\%$, $\sigma = 0.11\,\mu$, $\hat{a} = 24\,\mu$m; (**b**) J_x^*, $\mu = 1.05\%$, $\sigma = 0.19\,\mu$, $\hat{a} = 24\,\mu$m; (**c**) J_y^*, $\mu = 0.9\%$, $\sigma = 0.11\mu$, $\hat{a} = 24\,\mu$m; (**d**) J_y^*, $\mu = 1.05\%$, $\sigma = 0.19\,\mu$, $\hat{a} = 24\,\mu$m; (**e**) J_z^*, $\mu = 0.9\%$, $\sigma = 0.11\,\mu$, $\hat{a} = 24$; (**f**) J_z^*, $\mu = 1.05\%$, $\sigma = 0.19\,\mu$, $\hat{a} = 24\,\mu$m.

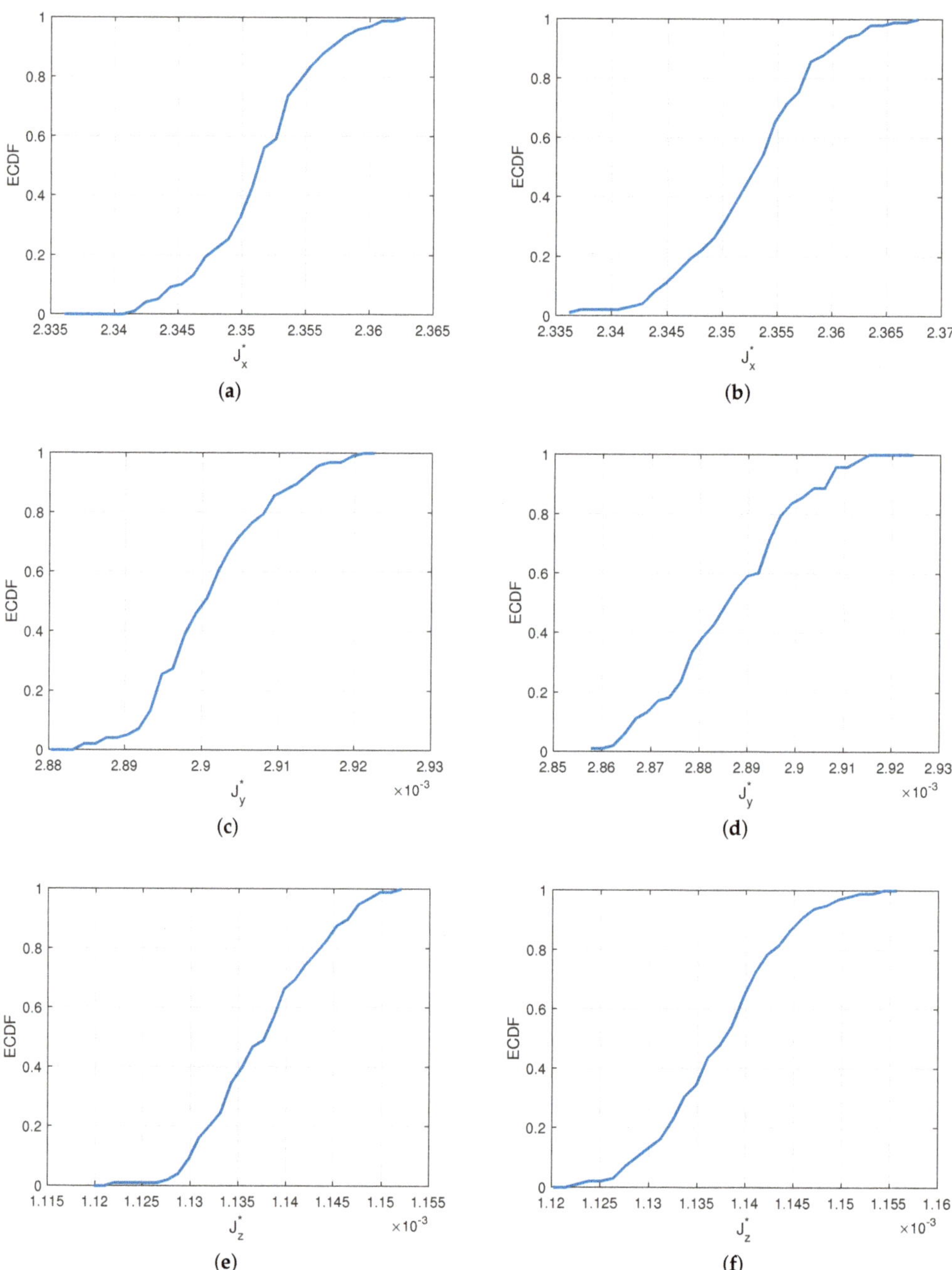

Figure 14. Identified probabilistic models (empirical cumulated distribution functions) for generating distributions of $\overline{\mathbf{J}}^*$ components at the macroscale. Values of $\overline{\mathbf{J}}^*$ are reported for a fixed value of macroboundary condition $\Phi_2 - \Phi_1 = 144$ V; (**a**) J_x^*, $\mu = 0.9\%$, $\sigma = 0.11\,\mu$, $\hat{a} = 24$ μm; (**b**) J_x^*, $\mu = 1.05\%$, $\sigma = 0.19\,\mu$, $\hat{a} = 24$ μm; (**c**) J_y^*, $\mu = 0.9\%$, $\sigma = 0.11\mu$, $\hat{a} = 24$ μm; (**d**) J_y^*, $\mu = 1.05\%$, $\sigma = 0.19\,\mu$, $\hat{a} = 24$ μm; (**e**) J_z^*, $\mu = 0.9\%$, $\sigma = 0.11\,\mu$, $\hat{a} = 24$; (**f**) J_z^*, $\mu = 1.05\%$, $\sigma = 0.19\,\mu$, $\hat{a} = 24$ μm.

7. Conclusions

A stochastic data-driven multilevel finite-element (FE^2) method was proposed to solve nonlinear heterogeneous structures with uncertainties at both the micro- and the macrolevel. A hybrid neural-network–interpolation (NN–I) scheme was developed to improve the accuracy of NN surrogate models, allowing for the use of a lower number of representative volume element (RVE) nonlinear calculations, which serve as a database to train the neural networks. This NN–I surrogate model was used to develop a data-driven method for nonlinear heterogeneous conduction in a stochastic framework: uncertainties can be included on both the micro- and the macrolevel. More specifically, the drastic reduction in computational costs brought by the NN-I surrogate model allows Monte Carlo simulations of nonlinear heterogeneous structures. This framework was applied to propagate uncertainties in such nonlinear multiscale models, and demonstrated that it can be used to identify probabilistic models related to some quantities of interest at the macroscale in a fully nonlinear, anisotropic context.

Author Contributions: Conceptualization, J.Y and V.P.; software, X.L. and L.P.; formal analysis, J.Y. and V.P.; investigation, X.L., J.Y., V.P., L.P. and I.K; writing—original draft preparation, J.Y., X.L. and V.P.; writing—review and editing, X.L., J.Y., V.P., L.P. and I.K.; supervision, J.Y.; project administration, J.Y.; funding acquisition, X.L. All authors have read and agreed to the published version of the manuscript.

Funding: Xiaoxin LU thanks the support from SIAT Innovation Program for Excellent Young Researchers 346 (E1G045).

Institutional Review Board Statement: Not applicable.

Informed Consent Statement: Not applicable.

Data Availability Statement: The data that support the findings of this study are available from the corresponding author upon reasonable request.

Acknowledgments: Xiaoxin Lu thanks the SIAT Innovation Program for Excellent Young Researchers (E1G045) for the support.

Conflicts of Interest: The authors declare no conflict of interest.

References

1. Liu, Y.; St-Pierre, L.; Fleck, N.A.; Deshpande, V.S.; Srivastava, A. High fracture toughness micro-architectured materials. *J. Mech. Phys. Solids* **2020**, *143*, 104060. [CrossRef]
2. Abueidda, D.W.; Bakir, M.; Al-Rub, R.K.A.; Bergström, J.S.; Sobh, A.N.; Jasiuk, I. Mechanical properties of 3D printed polymeric cellular materials with triply periodic minimal surface architectures. *Mater. Des.* **2017**, *122*, 255–267. [CrossRef]
3. Moini, M.; Olek, J.; Youngblood, J.P.; Magee, B.; Zavattieri, P.D. Additive manufacturing and performance of architectured cement-based materials. *Adv. Mater.* **2018**, *30*, 1802123. [CrossRef] [PubMed]
4. Kamkar, M.; Sultana, S.M.N.; Pawar, S.P.; Eshraghian, A.; Erfanian, E.; Sundararaj, U. The key role of processing in tuning nonlinear viscoelastic properties and microwave absorption in CNT-based polymer nanocomposites.*Mater. Today-Commun.* **2020**, *24*, 101010. [CrossRef]
5. Qi, X.-Y.; Yan, D.; Jiang, Z.; Cao, Y.-K.; Yu, Z.-Z.; Yavari, F.; Koratkar, N. Enhanced electrical conductivity in polystyrene nanocomposites at ultra-low graphene content.*ACS Appl. MAter. Int.* **2011**, *3*, 3130–3133. [CrossRef]
6. Milton, G.W. *Theory of Composites*; Cambridge University Press: Cambridge, UK, 2002; ISBN 9780521781251.
7. Torquato, S. *Random Heterogeneous Materials: Microstructure and Macroscopic Properties*; Springer Science and Business Media: New York, NY, USA, 2001; ISBN 978-1-4757-6355-3.
8. Yvonnet, J. *Computational Homogenization of Heterogeneous Materials with Finite Elements*; Springer: Cham, Switzerland, 2019; ISBN 978-3-030-18382-0.
9. Feyel, F. Multiscale FE^2 elastoviscoplastic analysis of composite structure. *Comput. Mater. Sci.* **1999**, *16*, 433–454. [CrossRef]
10. Feyel, F.; Chaboche, J.L. FE^2 multiscale approach for modelling the elastoviscoplastic behaviour of long fibre SiC/Ti composite materials. *Comput. Methods Appl. Mech. Eng.* **2000**, *183*, 309–330. [CrossRef]
11. Ozdemir, I.; Brekelmans, W.A.M.; Geers, M.G.D. Computational homogenization for heat conduction in heterogeneous solids. *Int. J. Numer. Methods Eng.* **2008**, *73*,185–204. [CrossRef]
12. Geers, M.G.D.; Kouznetsova, V.G.; Brekelmans, W.A.M. Multi-scale computational homogenization: Trends and challenges. *J. Comput. Appl. Math.* **2010**, *234*, 2175–2182. [CrossRef]

13. Terada, K.; Kikuchi, N. A class of general algorithms for multi-scale analysis of heterogeneous media. *Comput. Methods Appl. Mech. Eng.* **2001**, *190*, 5427–5464. [CrossRef]

14. Kouznetsova, V.G.; Geers, M.G.D.; Brekelmans, W.A.M. Multi-scale constitutive modeling of heterogeneous materials with gradient enhanced computational homogenization scheme. *Int. J. Numer. Methods Eng.* **2002**, *54*, 1235–1260. [CrossRef]

15. Ghosh, S.; Lee, K.; Raghavan, P. A multilevel computational model for multi-scale damage analysis in composite and porous media. *Int. J. Solids Struct.* **2001**, *38*, 2335–2385. [CrossRef]

16. Geers, M.G.D.; Yvonnet, J. Multiscale modeling of microstructure-property relations.*MRS Bull.* **2016**, *41*, 610–616. [CrossRef]

17. Yvonnet, J.; He, Q.-C. The reduced model multiscale method (R3M) for the non-linear homogenization of hyperelastic media at finite strains. *J. Comput. Phys.* **2007**, *223*, 341–368. [CrossRef]

18. Covezzi, F.; de Miranda, S.; Fritzen, F.; Marfia, S.; Sacco, E. Comparison of reduced order homogenization techniques: Prbmor, nutfa and mxtfa. *Meccanica* **2018**, *53*, 1291–1312. [CrossRef]

19. Michel, J.C.; Moulinec,H.; Suquet, P. Effective properties of composite materials with periodic microstructure: A computational approach. *Comput. Methods Appl. Mech. Eng.* **1999**, *172*, 109–143. [CrossRef]

20. Fang, G.; Wang, B.; Liang, J. A coupled Fe-FFT multiscale method for progressive damage analysis of 3D braided composite beam under bending load. *Compos. Sci. Technol.* **2019**, *181*, 107691. [CrossRef]

21. van Tuijl, A.R.; Remmers, J.J.C.; Geers, M.G.D. Multi-dimensional wavelet reduction for the homogenisation of microstructures. *Comput. Methods Appl. Mech. Eng.* **2020**, *359*, 112652. [CrossRef]

22. Roussette, S.; Michel, J.C.; Suquet, P. Non uniform transformation field analysis of elastic-viscoplastic composites. *Compos. Sci. Technol.* **2009**, *69*, 22–27. [CrossRef]

23. Liu, Z.; Bessa, M.A.; Liu, W.K. Self-consistent clustering analysis: An efficient multi-scale scheme for inelastic heterogeneous materials. *Comput. Methods Appl. Mech. Eng.* **2016**, *306*, 319–341. [CrossRef]

24. Bessa, M.A.; Bostanabad, R.; Liu, Z.; Hu, A.; Apley, D.W.; Brinson, C.; Chen, W.; Liu, W.K. A framework for data-driven analysis of materials under uncertainty: Countering the curse of dimensionality. *Comput. Methods Appl. Mech. Eng.* **2017**, *320*, 633–667. [CrossRef]

25. Fritzen, F.; Hodapp, M.; Leuschner, M. Gpu accelerated computational homogenization based on a variational approach in a reduced basis framework. *Comput. Methods Appl. Mech. Eng.* **2014**, *278*, 186–217. [CrossRef]

26. He, W.; Avery,P.; Farhat, C. In situ adaptive reduction of nonlinear multiscale structural dynamics models. *Int. J. Numer. Methods Eng.* **2020**, *121*, 4971–4988. [CrossRef]

27. Yvonnet, J.; Gonzalez, D.; He, Q.-C. Numerically explicit potentials for the homogenization of nonlinear elastic heterogeneous materials. *Comput. Methods Appl. Mech. Eng.* **2009**, *198*, 2723–2737. [CrossRef]

28. Le, B.A.; Yvonnet, J.; He, Q.-C. Computational homogenization of nonlinear elastic materials using neural networks. *Int. J. Numer. Methods Eng.* **2015**, *104*, 1061–1084. [CrossRef]

29. Fritzen, F.; Kunc, O. Two-stage data-driven homogenization for nonlinear solids using a reduced order model. *Eur. J. Mech. A/Solids* **2018**, *69*, 201–220. [CrossRef]

30. Lu, X.; Giovanis, D.; Yvonnet, J.; Papadopoulos, V.; Detrez, F.; Bai, J. A data-driven computational homogenization method based on neural networks for the nonlinear anisotropic electrical response of graphene/polymer nanocomposites. *Comput. Mech.* **2019**, *64*, 307–321. [CrossRef]

31. Li, B.; Zhuang, X.Y. Multiscale computation on feedforward neural network and recurrent neural network. *Front. Struct. Civ. Eng.* **2020**, *14*, 1285–1298. [CrossRef]

32. Logarzo, H.J.; Capuano, G.; Rimoli, J.J. Smart constitutive laws: Inelastic homogenization through machine learning. *Comput. Methods Appl. Mech. Eng.* **2021**, *373*, 113482. [CrossRef]

33. Nguyen-Thanh, V.M.; Nguyen, L.T.K.; Rabczuk, T.; Zhuang, X. A surrogate model for computational homogenization of elastostatics at finite strain using high-dimensional model representation-based neural network. *Int. J. Numer. Methods Eng.* **2020**, *121*, 4811–4842. [CrossRef]

34. Wang, K.; Sun, W. A multiscale multi-permeability poroplasticity model linked by recursive homogenizations and deep learning. *Comput. Methods Appl. Mech. Eng.* **2018**, *334*,337–380. [CrossRef]

35. Ghavamian, F.; Simone, A. Accelerating multiscale finite element simulations of history-dependent materials using a recurrent neural network. *Comput. Methods Appl. Mech. Eng.* **2019**, *357*,112594. [CrossRef]

36. Wu, L.; Zulueta, K.; Major, Z.; Arriaga, A.; Noels, L. Bayesian inference of non-linear multiscale model parameters accelerated by a deep neural network. *Comput. Methods Appl. Mech. Eng.* **2020**, *360*,112693. [CrossRef]

37. Peigney, M. A fourier-based machine learning technique with application in engineering. *Int. J. Numer. Methods Eng.* **2021**, *122*, 866–897. [CrossRef]

38. Rocha, I.B.C.M.; Kerfriden, P.; van der Meer, F.P. On-the-fly construction of surrogate constitutive models for concurrent multiscale mechanical analysis through probabilistic machine learning. *J. Comput. Phys. X* **2021**, *9*,100083. [CrossRef]

39. Rocha, I.B.C.M.; Kerfriden, P.; van der Meer, F.P. Micromechanics-based surrogate models for the response of composites: A critical comparison between a classical mesoscale constitutive model, hyper-reduction and neural networks. *Eur. J. -Mech. A/Solids***2020**, *82*, 103995. [CrossRef]

40. Avery, P.; Huang, D.Z.; He, W.; Ehlers, J.; Derkevorkian, A.; Farhat, C. A computationally tractable framework for nonlinear dynamic multiscale modeling of membrane woven fabrics. *Int. J. Numer. Methods Eng.* **2021**, *122*, 2598–2625. [CrossRef]

41. Wu, L.; Nguyen, V.D.; Kilingar, N.G.; Noels, L. A recurrent neural network-accelerated multi-scale model for elasto-plastic heterogeneous materials subjected to random cyclic and non-proportional loading paths. *Comput. Methods Appl. Mech. Eng.* **2020**, *369*, 113234. [CrossRef]
42. Mozaffar, M.; Bostanabad, R.; Chen, W.; Ehmann, K.; Cao, J.; Bessa, M.A. Deep learning predicts path-dependent plasticity. *Proc. Natl. Acad. Sci. USA* **2019**, *116*, 26414–26420. [CrossRef]
43. Bhattacharjee, S.; Matouš, K. A nonlinear manifold-based reduced order model for multiscale analysis of heterogeneous hyperelastic materials. *J. Comput. Phys.* **2016**, *313*, 635–653. [CrossRef]
44. Bhattacharjee, S.; Matouš, K. A nonlinear data-driven reduced order model for computational homogenization with physics/pattern-guided sampling. *Comput. Methods Appl. Mech. Eng.* **2020**, *359*, 112657. [CrossRef]
45. Clément, A.; Soize, C.; Yvonnet, J. Computational nonlinear stochastic homogenization using a non-concurrent multiscale approach for hyperelastic heterogenous microstructures analysis. *Int. J. Numer. Methods Eng.* **2012**, *91*, 799–824. [CrossRef]
46. Clément, A.; Soize, C.; Yvonnet, J. Uncertainty quantification in computational stochastic multiscale analysis of nonlinear elastic materials. *Comput. Methods Appl. Mech. Eng.* **2013**, *254*, 61–82. [CrossRef]
47. Rao, C.; Liu, Y. Three-dimensional convolutional neural network (3D-CNN) for heterogeneous material homogenization. *Comput. Mater. Sci.* **2020**, *184*, 109850. [CrossRef]
48. Lu, X.; Yvonnet, J.; Detrez, F.; Bai, J. Multiscale modeling of nonlinear electric conductivity in graphene-reinforced nanocomposites taking into account tunnelling effect. *J. Comput. Phys.* **2017**, *337*, 116–131. [CrossRef]
49. Liu, M.; Papageorgiou D.G.; Li, S.; Lin, K.; Kinloch, I.A.; Young, R.J. Micromechanics of reinforcement of a graphene-based thermoplastic elastomer nanocomposite. *Compos. Part A Appl. Sci. Manuf.* **2018**, *110*, 84–92. [CrossRef]
50. Yvonnet, J.; He, Q.-C.; Toulemonde, C. Numerical modelling of the effective conductivities of composites with arbitrarily shaped inclusions and highly conducting interface. *Compos. Sci. Technol.* **2008**, *68*, 2818–2825. [CrossRef]
51. Simmons, J.G. Electric tunnel effect between dissimilar electrodes separated by a thin insulating film. *J. Appl. Phys.* **1963**, *34*, 2581–2590. [CrossRef]
52. Eglajs, V.; Audze, P. New approach to the design of multifactor experiments. *Probl. Dyn. Strengths* **1977**, *35*, 104–107.
53. Ghanem, G.G.; Spanos, P.D. *Stochastic Finite Elements: A Spectral Approach*; Dover Publications: New York, NY, USA, 2003; ISBN 0-486-42818-4.
54. van der Vaart, A.W. *Asymptotic Statistics*; Cambridge University Press: Cambridge, UK, 1998; ISBN 0521784506.

Article

A Data-Driven Learning Method for Constitutive Modeling: Application to Vascular Hyperelastic Soft Tissues

David González [1] , Alberto García-González [2], Francisco Chinesta [3] and Elías Cueto [1,*]

[1] Aragon Institute of Engineering Research, Universidad de Zaragoza, 50018 Zaragoza, Spain; gonzal@unizar.es

[2] Laboratori de Càlcul Numèric, E.T.S. de Ingeniería de Caminos, Universitat Politècnica de Catalunya, 08034 Barcelona, Spain; berto.garcia@upc.edu

[3] ESI Group chair, Processes and Engineering in Mechanics and Materials (PIMM) Laboratory, Arts et Metiers Institute of Technology, 75013 Paris, France; Francisco.Chinesta@ensam.eu

* Correspondence: ecueto@unizar.es

Received: 21 April 2020; Accepted: 15 May 2020 ; Published: 18 May 2020

Abstract: We address the problem of machine learning of constitutive laws when large experimental deviations are present. This is particularly important in soft living tissue modeling, for instance, where large patient-dependent data is found. We focus on two aspects that complicate the problem, namely, the presence of an important dispersion in the experimental results and the need for a rigorous compliance to thermodynamic settings. To address these difficulties, we propose to use, respectively, Topological Data Analysis techniques and a regression over the so-called General Equation for the Nonequilibrium Reversible-Irreversible Coupling (GENERIC) formalism (M. Grmela and H. Ch. Oettinger, Dynamics and thermodynamics of complex fluids. I. Development of a general formalism. Phys. Rev. E 56, 6620, 1997). This allows us, on one hand, to unveil the true "shape" of the data and, on the other, to guarantee the fulfillment of basic principles such as the conservation of energy and the production of entropy as a consequence of viscous dissipation. Examples are provided over pseudo-experimental and experimental data that demonstrate the feasibility of the proposed approach.

Keywords: machine learning; manifold learning; topological data analysis; GENERIC; soft living tissues; hyperelasticity; computational modeling

---

1. Introduction

Computational (bio-)mechanics is not absent of the data "fever". Even if the so-called data-driven computational mechanics discipline presents distinctive features over what is commonly known as "big data", the possibility of employing raw experimental data to perform simulations has attracted the attention of many researchers recently [1–9]. In these approaches, basic equations—those with a higher epistemic value, such as equilibrium or compatibility—are kept, while those which are frequently phenomenological in nature—constitutive equations—are substituted by experimental data, and hence, the name of this family of methods. In a related approach, the so-called *equation-free* approach substitutes the constitutive law of a material not by experimental data, but by a pseudo-experimental result coming from microscale, first-principle computations [10].

This approach allows us to avoid complex, costly-and often unsuccessful-parameter fitting to obtain the precise form of a constitutive equation. On the other hand, widely accepted constitutive equations have been built under rigorous standards that give rise to the automatic satisfaction of first principles. This is not evident for data-driven approaches.

The presence of noise, or large variance among experiments in the data, somehow complicates the task of obtaining results that do not violate the laws of thermodynamics. Previous works in the field address the presence of noise in the results, but do not guarantee the satisfaction of basic thermodynamic principles [11,12]. This effect is particularly important in the field of bioengineering, where soft living tissues present a very important dispersion of the obtained parameter values from sample to sample, see for instance [13,14] and references therein.

Another important issue is that of the highly nonlinear anisotropic behavior of soft living tissues, which are able to show besides, inelastic and viscoelastic responses. Many works have been published in the last decades. In this regard, these behaviors have been traditionally modeled and characterized as (possibly visco-)hyperelastic anisotropic materials by means of strain energy density functions, see for example, in [15–20] and references herein. Inelasticity in the data-driven setting has been analyzed in a number of previous works [21–23]. Even the more daring approaches employ deep learning to relate medical images to mechanical properties of the tissue [24]. In none of the mentioned works is a study made by which the thermodynamic consistency of the obtained results is guaranteed. If we combine this possibility with the presence of noise in the data, the chance of violating first principles comes into play.

Different approaches exist to guarantee a rigorous thermodynamic compliance. For instance, Raissi et al. developed the so-called *physics-informed deep learning* method for the solution of partial differential equations [25]. A similar approach has also been developed to model turbulence [26] or, in general, the generalized Langevin equation [27]. The authors have recently developed an alternative approach based on the so-called General Equation for the Nonequilibrium Reversible-Irreversible Coupling (GENERIC) [28–30]. The GENERIC equation, which will be described next, constitutes a generalization of the Hamiltonian description of physical systems under nonequilibrium settings. The employ of the GENERIC formalism thus guarantees the correct fulfillment of the first and second principles of thermodynamics, and gives rise to a machine learning method valid for the description of the system at any level, from the molecular dynamics governed by Newtonian laws to the invariant-based description at the thermodynamics level [31,32].

The proposed method can be seen as a particular instance of the continuous dynamical system approach to machine learning, first proposed by W. E and coworkers [33,34]. In the quest for a theoretical framework for deep learning, a parallelism has been established between deep neural networks (DNNs) and continuous dynamical systems. DNNs are thus viewed under this prism as a discretization of a continuous dynamical system that maps the set of observations to a nonlinear function that fits the data [33]. This alternative view opens the door to the enforcement of desired properties to the learning processes; for instance, to ask the resulting map to posses a Hamiltonian structure, among others. In this work, we force the resulting map, which is found by regression in a piecewise polynomial manner, to obey a GENERIC description. This will ensure, as mentioned earlier, the fulfillment of the first and second principles of thermodynamics. To show its strong potentiality in the field of mechanical modeling of biological tissues, the method thus developed will be applied as a proof of concept to the characterization of the passive mechanical behavior of porcine carotid tissue. This behavior turns out to be-under the experimental setting considered here-elastic, highly nonlinear, anisotropic at finite strains, and often modeled under the framework of hyperelasticity.

The outline of the paper is as follows: Section 2 details the proposed data-driven model and the developed tests for validation. In Section 2.1, we describe the proposed machine learning technique to fit the mechanical responses of biological materials. To fit the well-known hyperelastic response in soft living tissues, the thermodynamically consistent GENERIC approach (describing the physics of the problem) and the subsequent machine learning procedure (constitutive manifold) is detailed here. Next, since numerical fitting of such tissues is strongly subjected to dispersion and averaging (mainly provoked by experimental procedures, environmental issues during the manufacturing, and testing process), a treatment of the noise and dispersion by means of Topological Data Analysis (TDA) is proposed in Section 2.2. Our presented approach is tested upon both a pseudo-experimental data set and experimental tests, described in Sections 2.3 and 2.4, respectively. The numerical fitting results of both experiments are shown in Section 3. The paper ends in Section 4 with a discussion of the obtained results on the use of the proposed GENERIC-TDA methodology on the mechanical modeling of biological tissues.

2. Material and Methods

2.1. A GENERIC Approach to the Learning Procedure

Recently, W. E and coworkers established a very useful parallelism between DNNs and dynamical systems [33]. Consider a system governed by some state variables $z(z_0, t) : \mathcal{I} \to \mathcal{S}, z \in \mathcal{C}^1(0, T]$, such that its time evolution is established in the form

$$\frac{dz}{dt} = f(z, t), \quad z(0) = z_0 \tag{1}$$

where f is, in general, a nonlinear function—otherwise, the method is of little interest. $\mathcal{S}$ represents the phase space of the system (a set of judiciously chosen variables that both describe the energy of the system and can be measured experimentally). $\mathcal{I} = (0, T]$ represents the considered time interval.

For the selected time horizon T, the *flow* map

$$z_0 \to z(z_0, T)$$

is a nonlinear function of z. The dynamical system approach to supervised learning consists in determining f so that the resulting flow map is able to reproduce the experimental data. This parallelism offers the advantage of the vast knowledge developed so far in the field of dynamical systems, which could help us in developing both theoretical insights about DNNs and also to devise alternative routes for developing efficient learning strategies. DNNs can be thought of as discretized dynamical systems, such that every time step corresponds to a layer of the DNN.

In this work, we consider viscous-hyperelastic materials. Therefore, the learned function f must satisfy certain well-known principles dictated by thermodynamics. In the hyperelastic case, the right choice for f could arise from Hamiltonian mechanics, i.e.,

$$\frac{dz}{dt} = L(z)\nabla E(z) \tag{2}$$

where E represents the Hamiltonian or, in other words, the energy of the system. $L(z)$ represents the so-called Poisson matrix, a skew-symmetric matrix that depends on z, in general, but that results to be constant in many cases.

Should the system of interest not be conservative (or Hamiltonian), a new potential needs to be introduced in the formulation-entropy-giving rise to the GENERIC formalism [30]

$$\dot{z}_t = L(z_t)\nabla E(z_t) + M(z_t)\nabla S(z_t), \quad z(0) = z_0 \tag{3}$$

For Equation (3) to represent valid nonequilibrium thermodynamic processes, it must be supplemented with the so-called degeneracy conditions, i.e.,

$$L(z) \cdot \nabla S(z) = 0 \tag{4}$$

$$M(z) \cdot \nabla E(z) = 0 \tag{5}$$

If, as stated before, L is skew-symmetric; and choosing M to be symmetric, positive semidefinite, we obtain

$$\dot{E}(z) = \nabla E(z) \cdot \dot{z} = \nabla E(z) \cdot L(z)\nabla E(z) + \nabla E(z) \cdot M(z)\nabla S(z) = 0 \tag{6}$$

i.e., one ensures the conservation of energy in closed systems.

In turn,

$$\dot{S}(z) = \nabla s(z) \cdot \dot{z} = \nabla S(z) \cdot L(z)\nabla E(z) + \nabla S(z) \cdot M(z)\nabla S(z) \geq 0 \tag{7}$$

which is equivalent to the fulfillment of the second principle of thermodynamics. Thus, we notice how, by leveraging the dynamical systems equivalence, we efficiently enforce the conservation of energy and the production of entropy. For an in-depth discussion of the implications of the choice of phase space variables z, we refer the interested reader to our previous works in the field, [31,32]. In general, it is well-known that neither a particular choice of variables is needed, nor a particular scale for the description of the system at hand. The only need is that the variables in the phase space would be able to account for the energy of the system. That is why, in general, a Langevin equation is not suitable for this purpose, see [35] and references therein.

Following the dynamical systems equivalence, the next step consists in the determination, by regression from data, of the form of the elements of the GENERIC description of the dynamics, i.e., L, M, $\nabla E(z)$, and $\nabla S(z)$. To do so, assume a standard finite difference discretization of the time derivative,

$$\frac{z_{n+1} - z_n}{\Delta t} = \mathsf{L}(z_{n+1})\mathsf{DE}(z_{n+1}) + \mathsf{M}(z_{n+1})\mathsf{DS}(z_{n+1}) \tag{8}$$

where we denote $z_{n+1} = z_{t+\Delta t}$ and where L and M are the discrete version of the Poisson and friction operators, respectively. DE and DS represent the discrete gradients. In general, matrix L is constant over the process, while matrix M frequently varies.

While there exist machine learning techniques that are able to provide with the precise expressions of the terms involved in a particular PDE from data [2], we pursue a purely numerical route, in which we assume that the elements of the GENERIC formula have a particular manifold structure which is to be unveiled by our method. This is the concept of constitutive manifold that we first stated in our previous works [4,7,36].

The dynamical systems approach to the problem will thus consist of solving the following (possibly constrained) minimization problem within a time interval $\mathcal{J} \subseteq \mathcal{I}$:

$$\mu^* = \{\mathsf{L}, \mathsf{M}, \mathsf{DE}, \mathsf{DS}\} = \arg\min_{\mu} ||z(\mu) - z^{\mathrm{meas}}|| \tag{9}$$

with $z^{\mathrm{meas}} \subseteq Z$, a subset of the total available experimental results. $\mathsf{L}, \mathsf{M}, \mathsf{DE}, \mathsf{DS}$ will in general be dependent on z, and therefore the choice of size of z^{meas} (in other words, the number of regressions performed along the time interval, or the number of layers in the DNNs equivalent) will affect the accuracy of the method. An analysis of the influence of this choice was made in [31].

We finally approximate z by employing piecewise polynomials, so that gradient operators can be cast in matrix form, `a la finite elements, as

$$\mathsf{DE} = Az \tag{10}$$

$$\mathsf{DS} = Bz \tag{11}$$

where A and B represent the discrete, matrix form of the gradient operators leading to DE and DS, respectively.

The GENERIC description of a hyperelastic material is well known [37]. Indeed, for the Hamiltonian, conservative part of the constitutive equation, we have

$$z(x,t) = [x(X,t), p(X,t)]^{\top} \tag{12}$$

where $x = \phi(X) \in \Omega_t \subset \mathbb{R}^3$ represents the deformed configuration of the solid, Ω_t represents the deformed configuration of the solid at time t, and $p \in \mathbb{R}^3$ represents the material momentum density. In this case,

$$\dot{z} = \begin{bmatrix} \dot{x} \\ \dot{p} \end{bmatrix} = L\nabla E = \begin{bmatrix} 0_{3\times3} & I_{3\times3} \\ -I_{3\times3} & 0_{3\times3} \end{bmatrix} \begin{bmatrix} \frac{\partial E}{\partial x} \\ \frac{\partial E}{\partial p} \end{bmatrix} \tag{13}$$

The total energy of a hyperelastic body is known to be the sum

$$E = W + K \tag{14}$$

of elastic and kinetic energies. Here, we assume a strain energy density potential w of the form

$$W = \int_{\Omega_0} w(C)\, d\Omega \tag{15}$$

where Ω_0 represents the undeformed configuration of the solid and C represents the right Cauchy-Green deformation tensor. In a general isotropic case, the strain energy density would take the form $w = w(X, C, S)$. If the material under consideration is isotropic hyperelastic, we simply write $w = w(C)$. In turn, the kinetic energy will be

$$K = \int_{\Omega_0} \frac{1}{2\rho_0} |p|^2 \, d\Omega \tag{16}$$

Therefore, by numerically identifying the form of the energy potential, one readily observes that the conservative part of the usual constitutive hyperelastic law is found. If, in addition, the material shows a viscous dissipative behavior, the precise form of the entropy potential is to be found.

Once the learning method has been developed, we describe next how the experimental campaign was accomplished.

2.2. Treatment of Dispersion and Noise in Data

One of the most important aspects to consider when dealing with soft living tissues is the importance of the dispersion of experimental results. This will be highlighted in Section 3 below. The main objective of the proposed technique is the determination of a constitutive manifold for the term of the GENERIC description of the physical phenomena, this type of dispersion needs to be treated efficiently.

To this end, we suggest to employ Topological Data Analysis (TDA) [38,39]. Behind the proposed method is the assumption of the existence of a constitutive manifold, a concept that we first introduced in [4]. Our set of experimental measurements, in the most general case, is assumed to be composed by D-tuples ($D = 9$ for this particular case) of the type

$$S = \{z = (U, P) \in (\mathbb{R}^3 \times \mathbb{R}^6)\} \tag{17}$$

where U represents the (right) stretch tensor and P the first Piola–Kirchhoff stress tensor. These experimental values are assumed to form a constitutive manifold in a high-dimensional space, see Figure 1. These values are, however, noisy. TDA is nevertheless able to extract the underlying geometry of this manifold, which will later be embedded onto a low-, d-dimensional manifold for interpolation purposes.

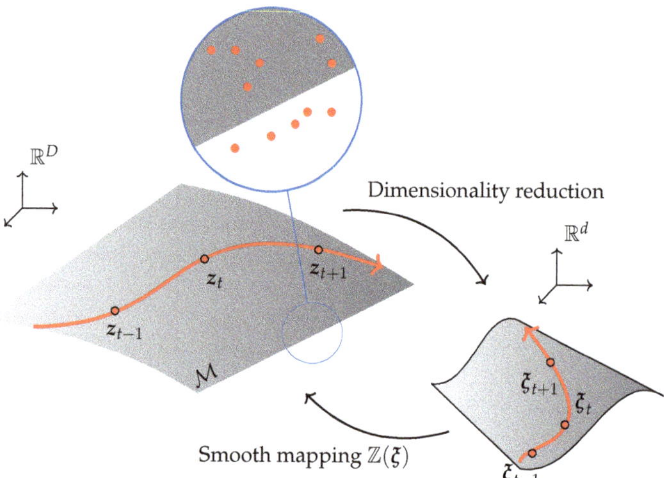

Figure 1. Hypothesis about the existence of a constitutive manifold in which the experimental results live. Despite the noise in the data, Topological Data Analysis (TDA) techniques will help us in unveiling the true geometry of the manifold in the high-dimensional setting, which will later be embedded onto a low dimensional space for the ease of computations.

TDA seeks to find the underlying topological structure of data. To this end, it employs a distance parameter R between experimental points. As we make R grow, points at distances lower than R will be connected by edges, triangles, and tetrahedra—in general, by k-simplexes, respectively, in one, two, three, ..., k dimensions. As simplexes appear, they form *simplicial complexes*. The study of the formation of this simplicial complex structure in the data is precisely the objective of TDA. The optimal R parameter that best describes the topology of the data is found by resorting to the so-called *persistence diagrams*.

In essence, simplicial homology reflects the number of holes in a given dimension for the simplicial complex. For instance, a chain of edges may close a hole, while the interior space within a tetrahedron is a void. If we record the precise value of R for which a hole or void (in any dimension) appears, and the R value for which it disappears—by the appearance of a filled triangle or tetrahedron, for instance—the resulting plots will indicate which topological features *persist* more. Those indicate the true topology of the data. For a graphical interpretation of this explanation, see Figure 2.

Under the TDA framework, noisy data produce topological structures with small persistence. The true topology (manifold structure) is described by these structures that persist the most. Once unveiled, the topological structure of data or, equivalently, the right shape of the data manifold, will allow us to interpolate data in the right topological space—in the tangent plane to the manifold at each datum—and not in Euclidean space.

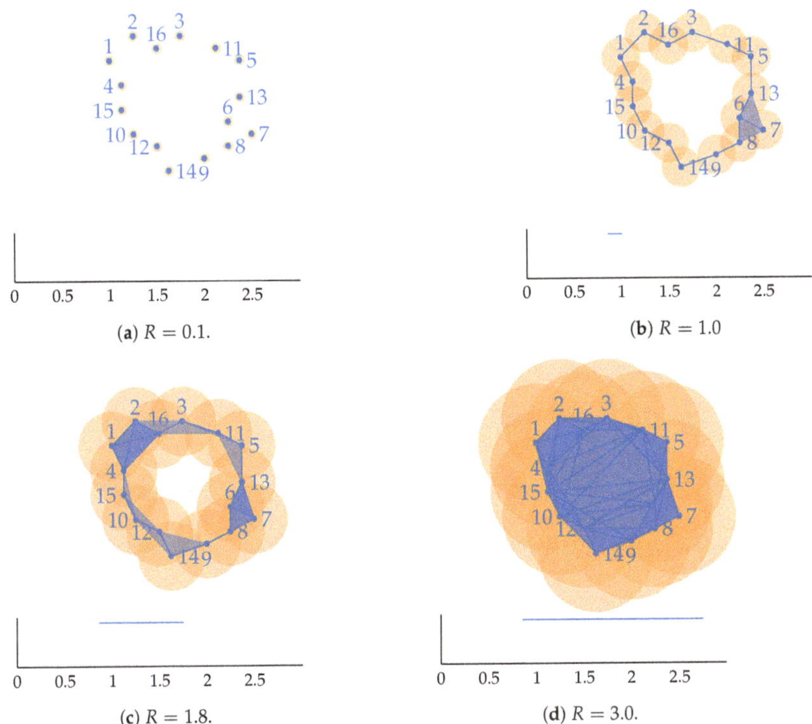

(a) $R = 0.1$.

(b) $R = 1.0$

(c) $R = 1.8$.

(d) $R = 3.0$.

Figure 2. Interpretation of TDA. Orange circles have diameter R, the topology parameter. Below each simplicial complex, the bar code corresponding to dimension 1 holes) is represented. (**a**) For a sufficiently small R parameter, say, 0.1, only a collection of data points is visible, with no topology at dimension 1. (**b**) Increasing R to 1.0 makes the first circular topology appear. This hole is visible for $R > 0.8$, hence the bar in the diagram from $R = 0.8$ to $R = 1.0$. (**c**) If we increase R, no perceptible changes are observed. Only one hole persists and it is reflected in the bar code below. (**d**) The hole disappears by formation of big triangles at about $R = 2.8$. From the observation of the bar code, we notice that the data set has the topology of a circle, with one single interior hole. In general, those holes or voids that persist the most reflect the persistent topology of the data.

2.3. Pseudo-Experimental Data—Learning a Visco-Hyperelastic Response

In this first example, we consider pseudo-experimental (numerical) data coming from a finite element simulation. We consider the same visco-hyperelastic material previously considered in [32], but this time altered with noise. This noise has a standard deviation of 10% of the mean value.

The considered material is assumed to obey a Mooney–Rivlin constitutive law

$$W = C_1(\bar{I}_1 - 3) + C_2(\bar{I}_2 - 3) + D_1(J - 1)^2 \tag{18}$$

with $\bar{I}_1 = J^{-\frac{2}{3}} I_1$ and $\bar{I}_2 = J^{-\frac{4}{3}} I_2$. $I_1 = \lambda_1^2 + \lambda_2^2 + \lambda_3^2$ and $I_2 = \lambda_1^2\lambda_2^2 + \lambda_2^2\lambda_3^2 + \lambda_3^2\lambda_1^2$ are the invariants of the right Cauchy-Green tensor C. In turn, J represents the determinant of the gradient of deformation tensor. We took $C_1 = 27.56$ MPa, $C_2 = 6.89$ MPa, and $D_1 = 0.0029$ MPa.

The viscous part of the behavior is modeled after a Prony series expansion of the type

$$\frac{G(t)}{G_0} = 1 - \sum_{i=1}^{2} \overline{g}_i^P \left(1 - \exp\left(-\frac{t}{\tau_i}\right)\right) \tag{19}$$

$$\frac{K(t)}{K_0} = 1 - \sum_{i=1}^{2} \overline{k}_i^P \left(1 - \exp\left(-\frac{t}{\tau_i}\right)\right) \tag{20}$$

with $\overline{g}_i^P = [0.2, 0.1]$ and $\overline{k}_i^P = [0.5, 0.2]$. The relaxation times take the values $\tau_i = [0.1, 0.2]$ seconds, respectively. Initial instantaneous Young's modulus and Poisson's ratio corresponding to these values are $E = 206.7$ MPa and $\nu = 0.45$, respectively.

With the material as just described, a plane-stress, biaxial experiment is reproduced. This experiment consists of two different loading steps, each one followed by a relaxation step. A total of 50 different data sets have been obtained for this experiment.

2.3.1. A mean GENERIC model.

A regression procedure is then accomplished for each one of the 50 different experiments, so as to determine their precise GENERIC expression. With the obtained values, we first compute the mean GENERIC model by simply taking mean values for each one of the GENERIC model components. This "mean" GENERIC model is compared to the noise-free numerical experiment, taken as ground truth.

2.3.2. Extracting the topology of data: GENERIC-TDA model.

Instead of just computing the mean values of each term of the GENERIC model, it seems judicious to employ TDA to unveil the topology of data and to determine the final GENERIC model by interpolating from the right neighboring experimental results. To this end, we considered a data set composed by 20 different loading states (all consisting of a load-relaxation-load-relaxation sequence) and the addition of noise (10% sdv) to each one of these processes, so as to obtain 50 different tests for each one of the 20 loading processes. This makes a total of one thousand different tests.

One of these tests (noise-free) is kept as the reference solution. We employ TDA to find the "neighboring" test to this reference solution and, by employing Kriging, to obtain the final numerical values of the GENERIC model. We would like to emphasize that interpolation is made not in the Euclidean space, but in the right tangent space to the data manifold, as unveiled by TDA techniques. Note that we speak of a tangent plane to a noisy manifold, since TDA is able to unveil the topology that persists the most, thus giving an apparent noise-free topology.

Weights for the interpolation are obtained by different Kriging interpolation techniques: Simple, Ordinary, and Local.

2.4. Learning the Constitutive Model of Porcine Carotid Tissue

One of the most intricate experimental procedures in the framework of solid mechanics is perhaps that of constitutive modeling of soft living tissues. Here, we will employ data previously obtained and presented in [14] for the constitutive modeling of porcine carotid tissue.

What is remarkable in soft living tissue modeling is, on one hand, the heterogeneity and anisotropy of the tissue; and on the other, the large differences between experimental values found in different specimens. For instance, the behavior of porcine carotid tissue—whose interest is to serve as a proxy of that of humans—differs strongly if the sample is extracted from proximal (i.e., close to the heart) positions of the vessel or if it is extracted from distal positions. Additionally, there is a strong anisotropy regarding circumferential versus longitudinal behavior.

In [14], a traditional fitting procedure was accomplished so as to determine the best fitting model for these data. Taking the mean of the experimental results arising from [14] as the only plausible reference solution—this is standard experimental procedure in the literature—what we did first was to determine by TDA the set of experimental results neighboring this (mean) reference solution. A GENERIC model was then determined for each one of these neighbor results. Three different approaches were then compared: i) a GENERIC model whose terms are computed as the mean values of each of the GENERIC models for each neighboring curve (the closest ones in the data manifold); and either ii) ordinary or iii) local Kriging interpolation techniques among these neighbors of the terms of a new GENERIC model. The result of our approach is a new GENERIC-TDA model whose integration in (pseudo-)time produces a prediction of the tissue behavior.

2.4.1. Experimental tests

To introduce to the reader the most significant details in the experimental models used for our numerical analysis, here we show a brief description of the sample's harvesting and tensile test protocols performed in [14]. The interested reader is referred to this article for a precise description of the experimental campaign.

We consider nine female pigs of 3.5 ± 0.45 months (mean $\pm$ SD). The experiments on these swine were approved by the Ethical Committee for Animal Research of the University of Zaragoza. All procedures were carried out in accordance with the Principles of Laboratory Animal Care (86/609/EEC Norm, incorporated into Spanish legislation through the RD 1021/2005).

For each one of the left and right carotids, proximal and distal regions were considered for mechanical testing, as mentioned before. At each location, circumferential and longitudinal strips, approximately 3-mm-wide and 11-mm-long, and 5-mm wide and 15-mm long, respectively, were cut. A total of 14 carotid specimens with 47 and 49 valid tests were performed for the proximal and distal zones, respectively. A minimum number of two test strips along each direction was accomplished for each specimen.

Simple tension tests of the carotid strips were performed in a high-precision-drive Instron Microtester 5548 system, see Figure 3. The procedure is properly described in [14], and consisted of the Instron Microtester 5548 System with two clamps holding the sample. The samples are subjected to the tensile test under a humidity-controlled environment to prevent sample drying.

The applied force was measured with a 5-N load cell with a minimal resolution of 0.001 N, and the axial strain was measured using a noncontact Instron 2663-281 video-extensometer equipped with a high-performance digital camera with a megapixel sensor ($0.5 \text{ m} \pm 0.5\%$)

Figure 3. Experimental setup. Instron Microtester 5548 System with two clamps holding the sample. The samples are subjected to the tensile test under a humidity-controlled environment to prevent sample drying.

Different loading and unloading cycles were applied that correspond to 60, 120, and 240 kPa (50%, 100%, and 200% of the estimated physiological stress state in the artery) at 30%/min of strain rate, which can be considered as quasi-static. Therefore, these experiments serve for the hyperelastic modeling of soft tissues, but not for their viscous characterization. The resulting GENERIC model will therefore be purely Hamiltonian under these conditions.

For each one of the fourteen carotids, proximal and distal measurements are obtained, thus giving a total of 28 datasets. Each dataset includes circumferential and longitudinal stresses and stretches, $z = \{\sigma_c, \lambda_c, \sigma_\ell, \lambda_\ell\}$. Due to the quasi-static nature of the experiments, time is actually a pseudo-time. In addition to these 14 stress–stretch curves for each location, a fifteenth curve is obtained by computing the mean value of the first 14. This curve will be taken as a sort of reference for comparison purposes, see Figure 4.

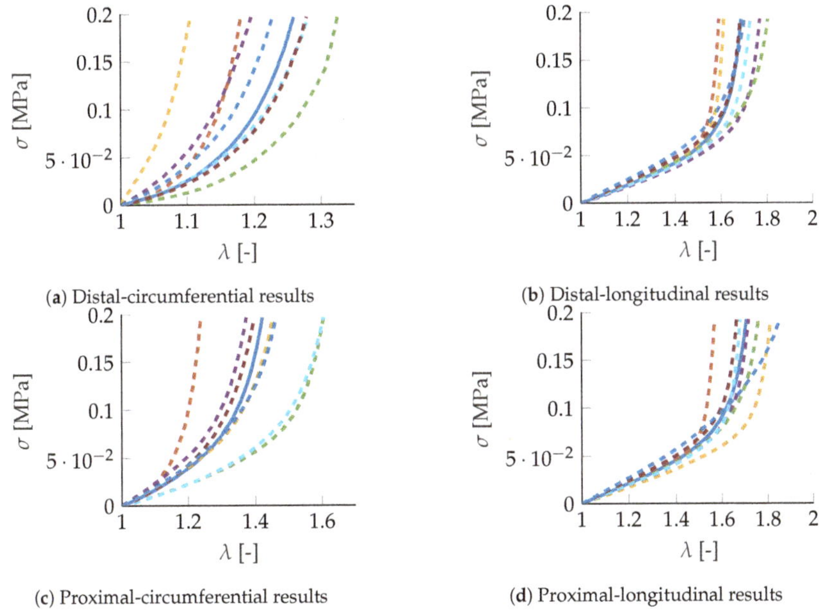

(a) Distal-circumferential results

(b) Distal-longitudinal results

(c) Proximal-circumferential results

(d) Proximal-longitudinal results

Figure 4. Experimental (**a**) distal-circumferential, (**b**) distal-longitudinal, (**c**) proximal-circumferential, and (**d**) proximal-longitudinal stress–stretch curves. The continuous blue line represents the mean values of the 14 experimental results, while the dashed lines represent the neighboring experiments, as found by TDA techniques.

3. Results

3.1. Numerical Fitting of the Pseudo-Experimental Data Set

Recalling Section 2.3, Figure 5 shows the "mean" GENERIC model when compared to the noise-free numerical experiment.

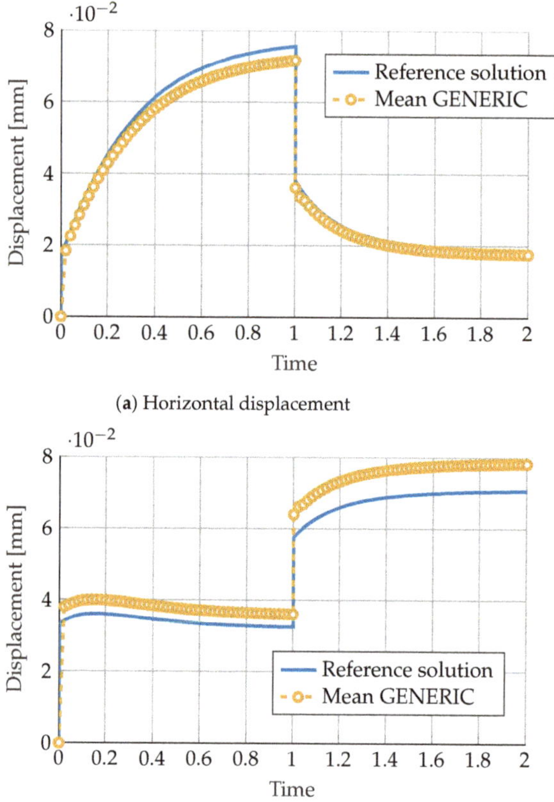

(**a**) Horizontal displacement

(**b**) Vertical displacement

Figure 5. Comparison of (**a**) horizontal and (**b**) vertical displacement predicted by a General Equation for the Nonequilibrium Reversible-Irreversible Coupling (GENERIC) model obtained as the mean of 50 different noisy GENERIC models. Comparison with the noise-free reference solution in continuous blue line.

As can be noticed from this figure, results show a poor accordance to the noise-free version of the data. Constructing a model by just computing the mean of each GENERIC model for noisy data seems not to be a good idea. If we consider it here, it is just because in the experimental framework, phenomenological models are very often obtained after computing means of the available results [14].

Figure 6 shows a displacement comparison among the noise-free sample and the full GENERIC-TDA model with different Kriging interpolation techniques. Additionally, Table 1 shows the obtained 2-norm errors of the mentioned model results.

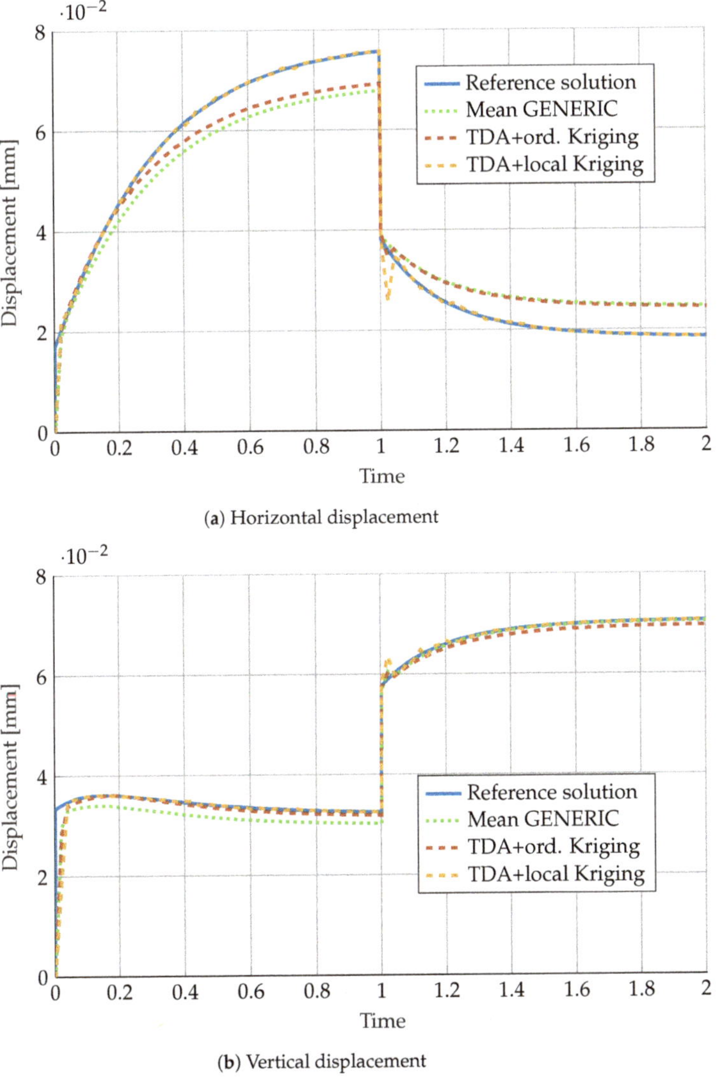

(**a**) Horizontal displacement

(**b**) Vertical displacement

Figure 6. Comparison of (**a**) horizontal and (**b**) vertical displacement predicted by a GENERIC model obtained as the mean of 50 different noisy GENERIC models. Comparison with the noise-free reference solution in continuous blue line and the solutions obtained by Kriging interpolation between neighbors predicted by Topological Data Analysis.

Table 1. 2-norm errors in the obtention of the GENERIC model.

Mean GENERIC	2.24%
Simple Kriging	16.46%
Ordinary Kriging	1.82%
Local Kriging	0.38%

It is worth noting the high degree of accuracy obtained by employing local Kriging procedures. In combination with TDA, this procedure is able not just to filter the artificial noise added to the data, but to provide a very accurate GENERIC model able to reproduce the visco-hyperelastic model from which pseudo-experimental data was obtained.

3.2. Numerical Fitting of Porcine Carotid Tissue

For the mean experimental curves (circumferential and longitudinal) of the distal samples, results are shown in Figure 7. It is worth noting that, in general, it seems not to be a good idea to just compute the mean values of the GENERIC models for each of the neighboring experimental curves. Particularly in the circumferential direction, the deviation from the reference solution is noteworthy. As in the previous section, results provided by Kriging interpolation outperform this approach. Particularly, local Kriging is found to provide the highest accuracy. The predicted behavior is almost indistinguishable from the mean experimental results.

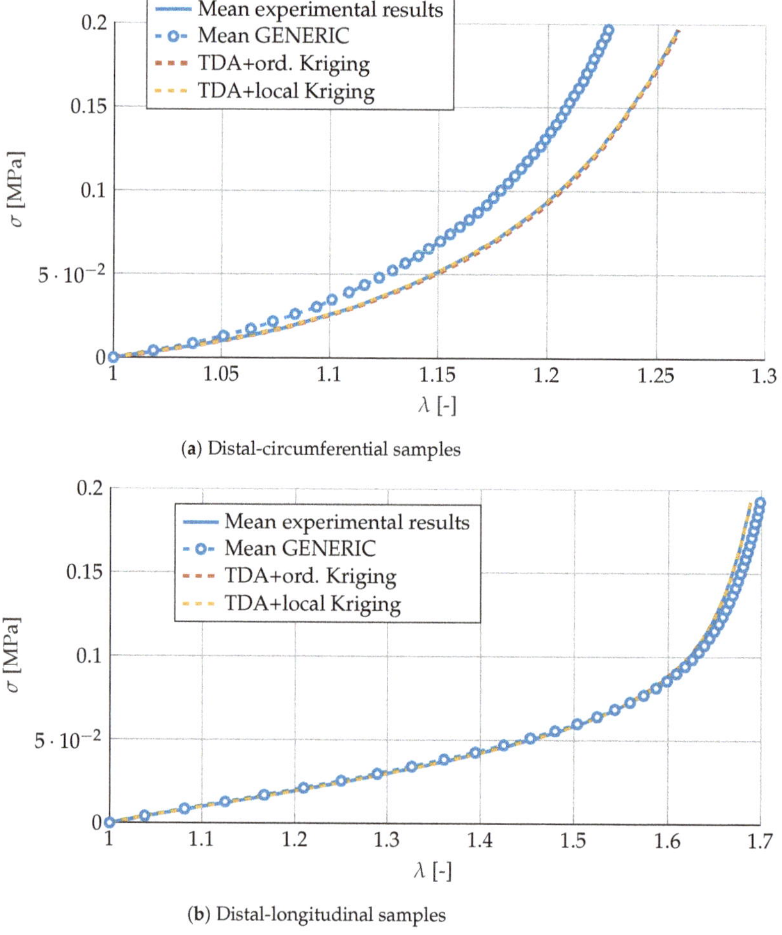

(**a**) Distal-circumferential samples

(**b**) Distal-longitudinal samples

Figure 7. Comparison of (**a**) distal-circumferential and (**b**) distal-longitudinal models predicted by mean GENERIC values, or by Kriging interpolation of those samples neighboring the reference solution.

With the weights just computed for the distal samples, we constructed a new GENERIC model for the proximal results. Its predictions are shown in Figure 8. Once again, by just computing the mean of the GENERIC terms for the neighboring curves does not seem to produce good results. However, with the Kriging weights computed for the distal samples, results for the proximal samples are equally good. This demonstrates the robustness of the proposed approach.

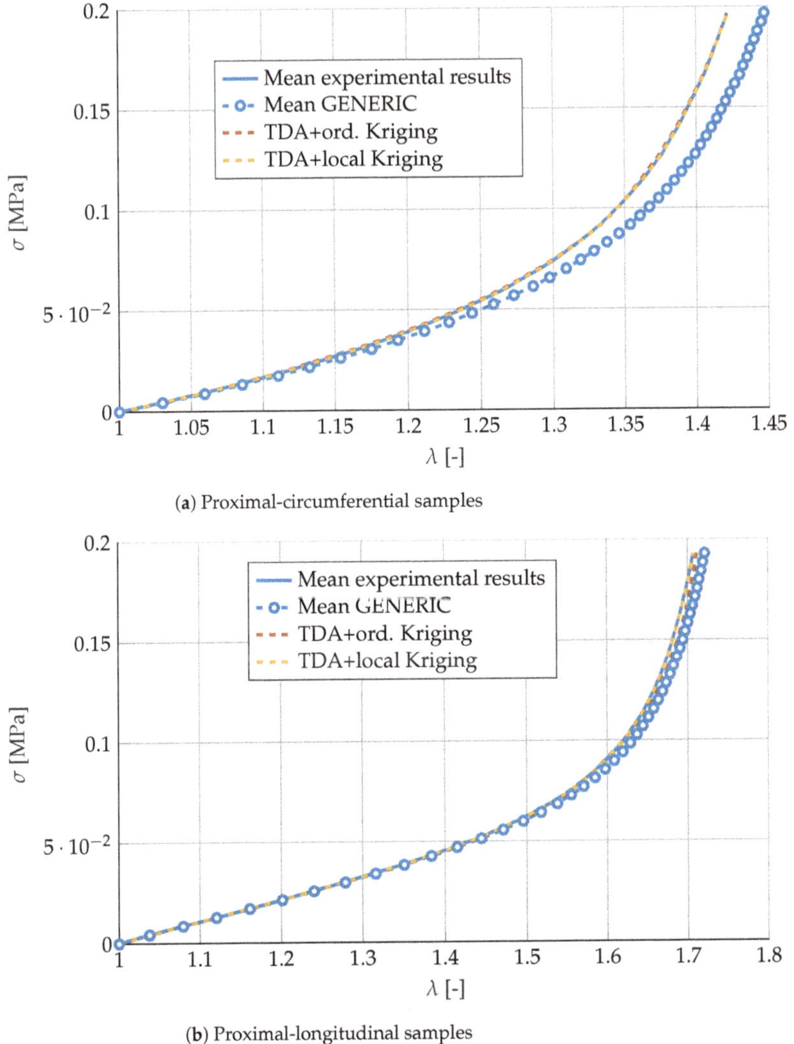

(**a**) Proximal-circumferential samples

(**b**) Proximal-longitudinal samples

Figure 8. Comparison of (**a**) proximal-circumferential and (**b**) proximal-longitudinal models predicted by mean GENERIC values, or by Kriging interpolation of those samples neighboring the reference solution.

It is worth noting that the computational cost of this procedure is by no means high: each sample took on average 2.52 seconds to obtain the corresponding GENERIC model.

4. Conclusions

In this paper, a new methodology for the data-driven learning of constitutive models is proposed. We made an emphasis on those cases in which large experimental deviations are present in the data. By first employing Topological Data Analysis techniques, we unveil the shape of the data manifold so as to allow us to perform interpolation on the right tangent plane to the manifold. Once the neighboring data are found, a GENERIC expression is found for the material under consideration. In other words, the precise form of the strain energy density and entropy potentials are found. This allows us to predict new loading states to a high degree of accuracy without the need to perform complex parameter fitting procedures to arrive to phenomenological models.

In addition, and in sharp contrast to other existing alternatives, our method is able to guarantee exact (to numerical precision) satisfaction of thermodynamic principles—conservation of energy and positive production of entropy—thanks to the GENERIC formalism.

To the best of our knowledge, no work has been performed in this line applied to biomedical living tissues. Despite the limitation of needing an admissible database to perform the learning process of our method, we strongly believe that the proposed GENERIC-TDA technique can be applied to the numerical fitting of highly nonlinear materials with sound accuracy, as shown in this manuscript. As a proof of concept, our results (developed in both synthetic and real experiments) show the high benefits of using data-driven models for materials simulation in fields where complex physical responses are present. We believe that machine learning methods combined with numerical modeling for biological systems (at any scale) is a very exciting young field with countless challenges and potential usefulness to both biomedical and numerical communities.

Author Contributions: Conceptualization, F.C. and E.C.; methodology, D.G., and E.C.; software, D.G. and A.G.-G.; validation, E.C. and F.C.; formal analysis, E.C. and F.C.; investigation, A.G.-G., D.G., and E.C.; data curation, A.G.-G.; writing—original draft preparation, E.C.; writing—review and editing, F.C. All authors have read and agreed to the published version of the manuscript.

Funding: This work has been supported by the Spanish Ministry of Economy and Competitiveness through Grant number DPI2017-85139-C2-1-R and by the Regional Government of Aragon and the European Social Fund, research group T24 20R. The support given by ESI Group to F.C. through the ESI Group Chair at ENSAM Paris and tho D.G. and E.C. through the project 2019-0060 "Simulated Reality" is also gratefully acknowledged.

Acknowledgments: The authors thank E. Peña for her helpful comments.

Conflicts of Interest: The authors declare no conflict of interest.

References

1. Kirchdoerfer, T.; Ortiz, M. Data-driven computational mechanics. *Comput. Methods Appl. Mech. Eng.* **2016**, *304*, 81–101, doi:10.1016/j.cma.2016.02.001.

2. Brunton, S.L.; Proctor, J.L.; Kutz, J.N. Discovering governing equations from data by sparse identification of nonlinear dynamical systems. *Proc. Natl. Acad. Sci.* **2016**, doi:10.1073/pnas.1517384113.

3. Liu, Z.; Bessa, M.; Liu, W.K. Self-consistent clustering analysis: an efficient multi-scale scheme for inelastic heterogeneous materials. *Comput. Methods Appl. Mech. Eng.* **2016**, *306*, 319–341.

4. Ibañez, R.; Borzacchiello, D.; Aguado, J.V.; Abisset-Chavanne, E.; Cueto, E.; Ladeveze, P.; Chinesta, F. Data-driven non-linear elasticity: constitutive manifold construction and problem discretization. *Comput. Mech.* **2017**, *60*, 813–826, doi:10.1007/s00466-017-1440-1.

5. Latorre, M.; Montáns, F.J. What-You-Prescribe-Is-What-You-Get orthotropic hyperelasticity. *Comput. Mech.* **2014**, *53*, 1279–1298, doi:10.1007/s00466-013-0971-3.

6. Bessa, M.; Bostanabad, R.; Liu, Z.; Hu, A.; Apley, D.W.; Brinson, C.; Chen, W.; Liu, W.K. A framework for data-driven analysis of materials under uncertainty: Countering the curse of dimensionality. *Comput. Methods Appl. Mech. Eng.* **2017**, *320*, 633–667.

7. Ibanez, R.; Abisset-Chavanne, E.; Aguado, J.V.; Gonzalez, D.; Cueto, E.; Chinesta, F. A manifold learning approach to data-driven computational elasticity and inelasticity. *Arch. Comput. Methods Eng.* **2018**, *25*, 47–57.

8. Latorre, M.; Peña, E.; Montáns, F.J. Determination and Finite Element Validation of the WYPIWYG Strain Energy of Superficial Fascia from Experimental Data. *Ann. Biomed. Eng.* **2017**, *45*, 799–810, doi:10.1007/s10439-016-1723-2.

9. Cilla, M.; Martinez, J.; Pena, E.; Martínez, M.Á. Machine Learning Techniques as a Helpful Tool Toward Determination of Plaque Vulnerability. *IEEE Trans. Biomed. Eng.* **2012**, *59*, 1155–1161, doi:10.1109/TBME.2012.2185495.

10. Kevrekidis, I.G.; Gear, C.W.; Hummer, G. Equation-free: The computer-aided analysis of complex multiscale systems. *AIChE J.* **2004**, *50*, 1346–1355.

11. Kirchdoerfer, T.; Ortiz, M. Data Driven Computing with noisy material data sets. *Comput. Methods Appl. Mech. Eng.* **2017**, *326*, 622–641, doi:10.1016/j.cma.2017.07.039.

12. Ayensa-Jiménez, J.; Doweidar, M.H.; Sanz-Herrera, J.A.; Doblaré, M. A new reliability-based data-driven approach for noisy experimental data with physical constraints. *Comput. Methods Appl. Mech. Eng.* **2018**, *328*, 752–774, doi:10.1016/j.cma.2017.08.027.

13. García, A.; Martínez, M.A.; Peña, E. Viscoelastic properties of the passive mechanical behavior of the porcine carotid artery: Influence of proximal and distal positions. *Biorheology* **2012**, *49*, 271–288.

14. García, A.; Peña, E.; Laborda, A.; Lostalé, F.; De Gregorio, M.; Doblaré, M.; Martínez, M. Experimental study and constitutive modelling of the passive mechanical properties of the porcine carotid artery and its relation to histological analysis: Implications in animal cardiovascular device trials. *Med. Eng. Phys.* **2011**, *33*, 665–676.

15. Holzapfel, G.A.; Gasser, T.C. A new constitutive framework for arterial wall mechanics and a comparative study of material models. *J. Elast.* **2000**, *61*, 1–48.

16. Pandolfi, A.; Holzapfel, G.A. Three-Dimensional Modeling and Computational Analysis of the Human Cornea Considering Distributed Collagen Fibril Orientations. *J. Biomech. Eng.* **2008**, *130*, 061006.

17. Peña, J.; Martínez, M.; Peña, E. A formulation to model the nonlinear viscoelastic properties of the vascular tissue. *Acta Mech.* **2011**, *217*, 63–74.

18. Holzapfel, G.A.; Sommer, G.; Gasser, C.T.; Regitnig, P. Determination of layer-specific mechanical properties of human coronary arteries with nonatherosclerotic intimal thickening and related constitutive modeling. *Am. J. Physiol.-Heart Circ. Physiol.* **2005**, *289*, H2048–H2058, doi:10.1152/ajpheart.00934.2004.

19. Gasser, T.C.; Ogden, R.W.; Holzapfel, G.A. Hyperelastic modelling of arterial layers with distributed collagen fibre orientations. *J. R. Soc. Interface* **2005**, *3*, 15–35.

20. Holzapfel, G.A.; Ogden, R.W. Constitutive modelling of arteries. *Proceedings of the Royal Society A: Mathematical, Physical and Engineering Sciences* **2010**, *466*, 1551–1597.

21. Ibañez, R.; Abisset-Chavanne, E.; Gonzalez, D.; Duval, J.; Cueto, E.; Chinesta, F. Hybrid constitutive modeling: Data-driven learning of corrections to plasticity models. *Int. J. Mater. Form.* **2018**, in press.

22. Eggersmann, R.; Kirchdoerfer, T.; Reese, S.; Stainier, L.; Ortiz, M. Model-free data-driven inelasticity. *Comput. Methods Appl. Mech. Eng.* **2019**, *350*, 81–99.

23. Zhang, M.; Benítez, J.M.; Montáns, F.J. Capturing yield surface evolution with a multilinear anisotropic kinematic hardening model. *Int. J. Solids Struct.* **2016**, *81*, 329–336, doi:10.1016/j.ijsolstr.2015.11.030.

24. Liang, L.; Liu, M.; Sun, W. A deep learning approach to estimate chemically-treated collagenous tissue nonlinear anisotropic stress-strain responses from microscopy images. *Acta Biomater.* **2017**, *63*, 227–235, doi:10.1016/j.actbio.2017.09.025.

25. Raissi, M.; Perdikaris, P.; Karniadakis, G.E. Physics Informed Deep Learning (Part I): Data-driven Solutions of Nonlinear Partial Differential Equations. *arXiv* **2017**, arXiv:1711.10561.

26. Wang, J.X.; Wu, J.L.; Xiao, H. Physics-informed machine learning approach for reconstructing Reynolds stress modeling discrepancies based on DNS data. *Phys. Rev. Fluids* **2017**, *2*, 034603.

27. Russo, A.; Durán-Olivencia, M.A.; Kevrekidis, I.G.; Kalliadasis, S. Deep learning as closure for irreversible processes: A data-driven generalized Langevin equation. *arXiv* **2019**, arXiv:1903.09562.

28. Grmela, M.; Öttinger, H.C. Dynamics and thermodynamics of complex fluids. I. Development of a general formalism. *Phys. Rev. E* **1997**, *56*, 6620–6632, doi:10.1103/PhysRevE.56.6620.

29. Öttinger, H.C. *Beyond Equilibrium Thermodynamics*; John Wiley Sons, Inc.: Hoboken, N.J., USA 2005.

30. Pavelka, M.; Klika, V.; Grmela, M. *Multiscale thermodynamics*; De Gruyter: Berlin, Germany, 2018.

31. González, D.; Chinesta, F.; Cueto, E. Thermodynamically consistent data-driven computational mechanics. 2018, submitted.

32. Gonzalez, D.; Chinesta, F.; Cueto, E. Learning corrections for hyperelastic models from data. *Front. Mater.* **2019**, *6*, 14.

33. Weinan, E. A Proposal on Machine Learning via Dynamical Systems. *Commun. Math. Stat.* **2017**, *5*, 1–11, doi:10.1007/s40304-017-0103-z.

34. Li, Q.; Chen, L.; Tai, C.; E, W. Maximum Principle Based Algorithms for Deep Learning. *J. Mach. Learn. Res.* **2018**, *18*, 1–29.

35. Español, P. Statistical Mechanics of Coarse-Graining. In *Novel Methods in Soft Matter Simulations*; Karttunen, M., Lukkarinen, A., Vattulainen, I., Eds.; Springer: Berlin/Heidelberg, Germany, 2004; pp. 69–115, doi:10.1007/978-3-540-39895-0_3.

36. Lopez, E.; Gonzalez, D.; Aguado, J.V.; Abisset-Chavanne, E.; Cueto, E.; Binetruy, C.; Chinesta, F. A Manifold Learning Approach for Integrated Computational Materials Engineering. *Arch. Comput. Methods Eng.* **2016**, 1–10, doi:10.1007/s11831-016-9172-5.

37. Romero, I. A characterization of conserved quantities in non-equilibrium thermodynamics. *Entropy* **2013**, *15*, 5580–5596.

38. Wasserman, L. Topological Data Analysis. *Annu. Rev. Stat. Its Appl.* **2018**, *5*, 501–532, doi:10.1146/annurev-statistics-031017-100045.

39. Munch, E. A User's Guide to Topological Data Analysis. *J. Learn. Anal.* **2017**, *4*, 47–61.

materials

Article

Hybrid Modelling by Machine Learning Corrections of Analytical Model Predictions towards High-Fidelity Simulation Solutions

Frederic E. Bock [1,*], Sören Keller [1], Norbert Huber [1] and Benjamin Klusemann [1,2]

1. Institute of Materials Mechanics, Helmholtz-Zentrum Hereon, 21502 Geesthacht, Germany; soeren.keller@hereon.de (S.K.); norbert.huber@hereon.de (N.H.); benjamin.klusemann@leuphana.de (B.K.)
2. Institute of Product and Process Innovation, Leuphana University of Lüneburg, 21335 Lüneburg, Germany
* Correspondence: frederic.bock@hereon.de

Abstract: Within the fields of materials mechanics, the consideration of physical laws in machine learning predictions besides the use of data can enable low prediction errors and robustness as opposed to predictions only based on data. On the one hand, exclusive utilization of fundamental physical relationships might show significant deviations in their predictions compared to reality, due to simplifications and assumptions. On the other hand, using only data and neglecting well-established physical laws can create the need for unreasonably large data sets that are required to exhibit low bias and are usually expensive to collect. However, fundamental but simplified physics in combination with a corrective model that compensates for possible deviations, e.g., to experimental data, can lead to physics-based predictions with low prediction errors, also despite scarce data. In this article, it is demonstrated that a hybrid model approach consisting of a physics-based model that is corrected via an artificial neural network represents an efficient prediction tool as opposed to a purely data-driven model. In particular, a semi-analytical model serves as an efficient low-fidelity model with noticeable prediction errors outside its calibration domain. An artificial neural network is used to correct the semi-analytical solution towards a desired reference solution provided by high-fidelity finite element simulations, while the efficiency of the semi-analytical model is maintained and the applicability range enhanced. We utilize residual stresses that are induced by laser shock peening as a use-case example. In addition, it is shown that non-unique relationships between model inputs and outputs lead to high prediction errors and the identification of salient input features via dimensionality analysis is highly beneficial to achieve low prediction errors. In a generalization task, predictions are also outside the process parameter space of the training region while remaining in the trained range of corrections. The corrective model predictions show substantially smaller errors than purely data-driven model predictions, which illustrates one of the benefits of the hybrid modelling approach. Ultimately, when the amount of samples in the data set is reduced, the generalization of the physics-related corrective model outperforms the purely data-driven model, which also demonstrates efficient applicability of the proposed hybrid modelling approach to problems where data is scarce.

Keywords: machine learning; analytical model; finite element model; artificial neural networks; model correction; feature engineering; physics based; data driven; laser shock peening; residual stresses

Citation: Bock, F.E.; Keller, S.; Huber, N.; Klusemann, B. Hybrid Modelling by Machine Learning Corrections of Analytical Model Predictions towards High-Fidelity Simulation Solutions. *Materials* **2021**, *14*, 1883. https://doi.org/10.3390/ma14081883

Academic Editor: Moncef L. Nehdi

Received: 11 March 2021
Accepted: 4 April 2021
Published: 10 April 2021

Publisher's Note: MDPI stays neutral with regard to jurisdictional claims in published maps and institutional affiliations.

1. Introduction

There is currently a surge in the application of machine learning algorithms in various fields of materials mechanics. In general, scientific and industrial research groups focus on the identification and utilization of one or more relationships along the process–structure–property–performance (p-s-p-p) chain [1]. In this domain, the application of machine learning techniques can be a key enabler for accelerated identification, characterization, understanding and optimization of processes, materials and parameters [2]. For instance, unique material descriptors can be qualified and quantified for material characterization [3–5]. Optimization and rapid design of novel manufacturing methods and involved materials [6,7] can be achieved, and inaccurate measurement techniques can be corrected [8]. The generation of knowledge and understanding to enable improved predictions of mechanical properties and performances, among others, can be acquired on the basis of experimental and/or numerical data in combination with machine learning models [9,10]. Furthermore, the integration of well-established physical laws into data-driven machine-learning models can be very beneficial to perform highly accurate predictions and inferences of involved phenomena [11,12]. However, besides these physics-informed machine learning methodologies, Chinesta et al. [13] introduced a hybrid modelling approach, where an efficient physics-based model shows some prediction errors that are corrected by a subsequent data-driven model to ultimately reach the anticipated solution.

Deployment of only either data-driven predictive models or calibrated physics-based models is accompanied with respective disadvantages based on each approach. Calibration of physics-based models can be difficult, expensive and time-costly even for domain experts, as it can be challenging or even impossible for physical quantities of interest to be accessible through experimental measurements. It is almost unattainable to represent the reality via such models only through data assimilation [14]. For purely data-driven approaches, the relevant relationships between input and output variables are required to be satisfactorily represented in the data set, as there is an absence of internal physics-related variables [15]. This creates the demand for a comprehensive database for the learning algorithm to represent those relationships. For problems that are still largely unknown, this can be a suitable approach; however, when some relations are already known, it is inefficient to create the need of a big-data-set for ensuring it represents all relevant aspects of the underlying physical laws that are required to be learned "from scratch" by the machine-learning algorithm [16]. In a study by Liu et al. (2020) [17], a data-driven surrogate model to predict the plane-strain stress intensity factor at the crack tip during fracture toughness tests is built with an adaptability and efficiency that is comparable to an analytical or empirical solution within their physical problem domains. In [17], high-fidelity numerical simulations are used to create the data-base for correlation of dimensionless inputs and outputs. However, due to the purely data driven approach, a vast number of computationally expensive simulation solutions are required for sufficient training of the surrogate model, which could create challenges for accuracy and generalization when switching to an experimental data source for training. Purely data-driven approaches can be beneficial for those problems where few relationships are identified, as they can help to detect hidden relationships in data; however, when established physical-laws apply and available data is scarse or biased, the utilization of physically-related data-driven approaches can be countervailing and utile [18,19].

Consequently, studies are focused on the aim to represent physical problems and their associated behaviour through physics-based models as well as on the pursuit to account for the deviation between those models and the reality via data-driven corrections. González et al. (2019) [20] performed corrections for hyperelastic models based on data-driven machine learning, whereas Ibáñez et al. (2018) [21] implemented a hybrid approach consisting of constitutive modelling and data-driven machine learning correction of plasticity models. In a manufacturing application example for metal forming production, Havinga et al. (2020) [22] performed real-time predictions via a hybrid modelling approach that contains physics-based simulations those predictive deviations to the real process are eliminated via an additional corrective model. Overall, the specific employment of machine learning models alongside governing physics-based relationships allows for highly valid predictions within materials mechanics and its related fields.

Generally, physics-based models might show prediction errors but as these deviations are systematic and not owed to noise, they can be accounted for separately. In combination, physics-based models and deviation models can be used to correctly predict a real system's behaviour. The advantages of using a calibrated model based on well-established physics, even when it shows deviations to reality, are that the compensating corrective model applied for achieving high prediction accuracy requires fewer samples and less complexity to approximate the deviation, since it is usually considerably less non-linear than the problem itself. This opens up the possibility to easily correct a physics-based model with a relatively simple correction model towards true/desired data points to assure an adequate representation of the behaviour by the system of interest [13]. Chupakhin et al. [8] introduced a corrective artificial neural network (ANN) for the hole drilling method, where residual stresses are determined based on measurements of elastic material behaviour, which are corrected towards the solution of a plasticity-including finite element (FE) model by an ANN. Thus, as opposed to correcting numerical models by empirical observations, in this case, biased experimental measurements can be successfully corrected through an ANN driven by physics-based numerical data.

The objective of this study is to build a hybrid model, consisting of a physics-based model and a data-driven corrective model, with low prediction errors even when training data is scarce. A semi-analytical model, originally proposed by Hu et al. [23], is employed as low-fidelity physics-based model, including a number of simplifications and a subsequent ANN is used to correct this solution towards a true reference solution provided by an FE model considered as high-fidelity. As example use-case, laser shock peening (LSP)-induced residual stress distributions over the specimen depth in aluminium alloy AA2024 are considered. In particular, since the representation of the relationships between residual stress distributions in dependence of LSP-generated pressure pulses over time is severely simplified in its semi-analytical model solution, we aim for the complementary corrective approach. Ultimately, high-fidelity approximation of the desired system behaviour is achieved by combining semi-analytical and ANN-correction models, which are both computationally efficient. In addition, when the data used for training, validation and testing is reduced, the predictions obtained via this hybrid modelling approach exhibit less errors than a purely data driven model. We propose a hybrid process model consisting of data-driven correction-learning of an LSP process model, which also shows good generalization ability, even when the parameter space of the training region is expanded and the available data becomes scarce.

2. Methods and Materials

The implemented corrective approach combines a semi-analytical model, which exhibits significant deviations in predictions outside its calibration parameter space, with a data-driven machine learning model correcting those deviations towards the solution of the high-fidelity model. The corrective model is required to be less complex, for solely representing a corrective component, compared to a purely data-driven prediction model mapping the more complex and complete relationships that are relevant. Additionally, this hybrid approach shows good generalization ability and also exhibits low prediction errors in an expanded input parameter space outside the parameter space used for training, as opposed to decreased generalization ability of a purely data-driven model, which is not physics-related. For the selected use-case of LSP, the residual stress distributions intended to be corrected are calculated via the semi-analytical model from Hu et al. [23]. An FE model was used for computing the desired reference residual stress distributions, which represent the true/desired data in this work. The correction task is developed through training, validating and testing of an ANN. Both numerical and semi-analytical models will be briefly introduced in the following two sections. For more details, the reader is referred to the original publications, as the focus of this study lies on the correction task where those models are assumed as black-box models and their detailed mechanisms are deliberately not intended to be relevant for the current study. (Note: the selected use-case LSP serves only as selected example. Generally speaking, the analytical model could be replaced by any physics-based model and the data from the FE model represents the corresponding, typically scarce, experimental data). Material parameters correspond to the aluminium alloy AA2024 in T3 heat treatment condition, frequently used in the aircraft industry for fuselage structures [24].

2.1. Laser Shock Peening

One of the main goals of the transportation industry is to reach weight, fuel and CO_2 savings as well as increase the sustainability of engineering components [25]. For improving the fatigue life of light-weight materials such as aluminium alloys, LSP has gained attention in scientific research and industrial application developments. LSP is known as residual stress modification technique to introduce high and deep compressive residual stresses in metallic components [26]. These compressive residual stresses can be used to enhance fatigue properties of metallic structures, which is of high interest for damage tolerant design concepts, as applied in aircraft structures. However, compressive residual stresses are always accompanied by fatigue-critical tensile residual stresses due to stress equilibrium. During LSP, short-time (nanosecond regime), high-energy (Joule regime) laser pulses are used to convert material at the surface into plasma. Plasma expansion initiates mechanical shock waves that cause local plastic strains in the material. After relaxation of the dynamic process, a characteristic residual stress field is developed, which contains both: Relatively high compressive residual stresses and balancing tensile residual stresses. Experimental process observation is very challenging and requires great effort due to the magnitudes of physical quantities, such as plasma pressure as well as temperature, and the short time scale. The knowledge of the residual stress fields is essential for efficient application of LSP, motivating the development of suitable prediction tools. Modelling of the LSP process is challenging due to the short time scale of the process, which, so far, leads to imprecise experimental determination of physical quantities occuring during shock wave propagation and plasma formation, such as material strain rates up to $10^6\,\mathrm{s}^{-1}$, plasma pressure of several GPa or the high plasma temperature; therefore, the utilized material model can exhibit determination inaccuracies regarding these quantities. There are various approaches to simulate the LSP process, such as FE models [27–29] or (semi) analytical models. While FE models represent the most commonly used modelling approach, to represent the three-dimensional physics involved in the complex LSP process, the considered semi-analytical model by Hu et al. [23] is computationally very efficient but does not provide any information on tensile stresses because stress equilibrium is neglected.

Other simplifications include the assumption of an infinite instead of finite specimen thickness as well as single value calculations of stresses at distinct model locations as opposed to averaged stress calculation based on extrapolation of finite element integration points towards nodes, among others. Since the considered LSP system and FE model uses quadratic pulse spots, see Keller et al. [29], the underlying assumption of a circular spot in the semi-analytical model represents a further simplification in the current case.

Ultimately, the proposed correction approach is employed to achieve low prediction errors while simultaneously using the implied physics and maintaining the computational efficiency of the analytical model. Such a hybrid modelling approach is new in the context of LSP, where the number of publications on the application of machine learning approaches for the LSP process is scarce, overall. Frija et al. [30] optimized the LSP surface conditions by using an FE model exposed to the laser-induced pressure pulse as well as Design of Experiments (DoE) to infer related laser parameters. They extended the work by the use of an ANN to efficiently predict significant characteristics of numerical compressive residual stress profiles and approximated a simplified 1st-order linear slope of residual stresses [31]. In this study, it is aimed for efficiently predicting the original non-linear distribution of compressive and tensile residual stresses, provided by an FE model, throughout the complete depth of the specimen. Wu et al. [32] also performed predictions of LSP-induced residual stresses via an ANN based on the laser profile and laser energy purely based on experimental data; thereby, not explicitly considering relevant physical relationships. Mathew et al. [33] used an ANN for the prediction and optimization of residual stress distributions induced by LSP, where the relative importance of four process parameters on residual stresses is investigated purely based on experimental data. In this work, the proposed hybrid model generates highly accurate predictions that are physics-related via the corrective approach of a physics-based analytical model.

2.2. Physical Models

In the following, the pressure pulse input definition for both physical models as well as the semi-analytical model and high-fidelity FE model, are described.

2.2.1. Pressure Pulse Definition for Physical Models

The definition of the pressure pulses over time, in Figure 1, is utilized as input for the semi-analytical model, see Figure 2a, and for the high-fidelity FE model, see Figure 3a. The pressure pulse over time is uniquely defined in this work based on three pressure pulse parameters: Maximum pressure P_{max}, the time of maximum pressure t_I and the pulse duration t_{II}, see Figure 1. This pressure pulse function is preferred in the utilized ABAQUS solver of the FE analysis since it is differentiable and assures efficiency and stability of the FE solver [34]. Note that the original semi-analytical model by Hu et al. [23] is slightly modified by using this pressure pulse as input, instead of laser parameters. Note: The pulse duration t_{II} is not considered in the semi-analytical model, as described in the following Section 2.2.2.

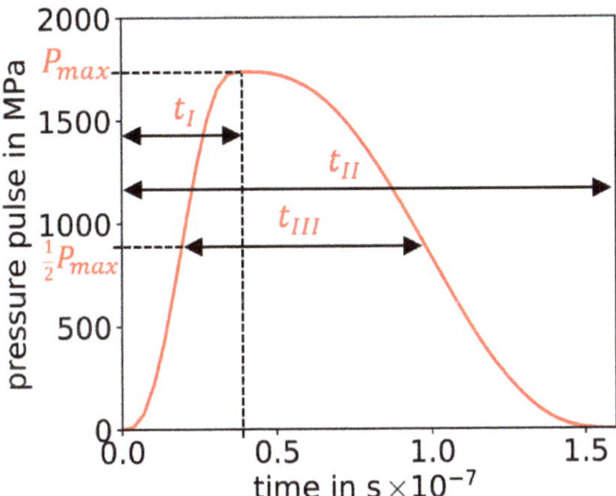

Figure 1. Pressure pulse over time including its uniquely defining parameters: Maximum pressure P_{max}, time of maximum pressure t_I and pulse duration t_{II}. As additional information, the full width at half maximum is given by t_{III}.

2.2.2. Low-Fidelity Model — Semi-Analytical Model

A semi-analytical LSP process model to predict residual stress profiles depending on the plasma pressure is developed by Hu et al. [23], which is adopted in this study. In the process model, a semi-infinite space and rotational symmetry are assumed since a circular laser focus is considered. Furthermore, single laser pulse impacts are modelled instead of a laser pulse sequence. The residual stress profile is evaluated along the symmetry axis. The LSP process of a single laser pulse impact is split into two phases: Loading and relaxation. During the loading phase, the pulse pressure from $t = 0$ to t_I is considered as input and during the relaxation phase, the resulting residual stresses are calculated (note that the pressure pulse interval from t_I to t_{II} is not considered in this model). Plasma induced stresses that are present during the loading phase are assumed to be superposed and fully developed stress fields that are caused by time dependent surface tractions of the plasma pressure, representing the elastic solution. The stress field caused by a single traction is described by closed-form expressions corresponding to the equation found for single forces, see Timoshenko and Goodier [35]. Plastic material deformation and resulting stresses are calculated by the McDowell Hybrid Algorithm [36]. A strain-rate dependent material model, including isotropic and kinematic hardening is employed. The strain-rate dependency of the yield stress is modelled by the Johnson–Cook model, where material parameters are listed in Table 1. After the application of the plasma pressure, the residual stress field is calculated during the relaxation phase; therefore, stresses are incrementally reduced while plastic deformation is taken into account to match stress and strain boundary conditions of an axisymmetric half space. A stress equilibrium is not calculated by this algorithm, as opposed to the FE analysis, which is explained in the following Section 2.2.3. For more details on the semi-analytical model, the interested reader is referred to the original work by Hu et al. [23]. Overall, the main involved physical phenomena are considered in the semi-analytical model but to a substantially simplified extent leading to a relatively narrow parameter space, where in combination with a subsequent correction, the desired high fidelity solution of the FE model within a much wider parameter space can be reached, nevertheless.

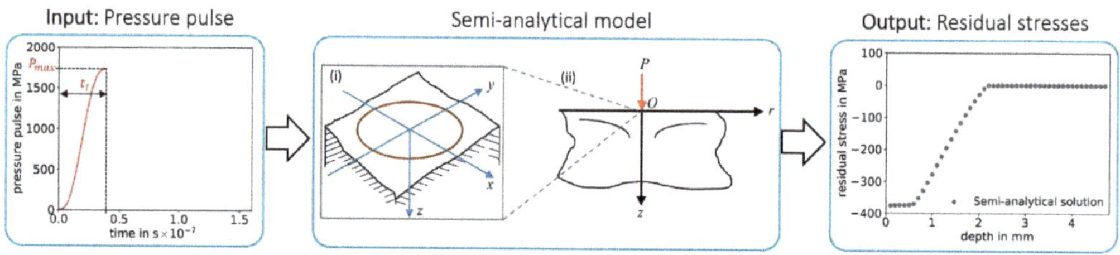

Figure 2. Illustration of the semi-analytical model by Hu et al. [23] for computing residual stresses induced by pressure pulse from Figure 1. Circular pressure pulse area (i) (in red) on the half-space model, which is simplified in (ii) as a concentrated normal load (in red) in the axisymmetric half-space model. Figures (i) and (ii) are republished with permission of the American Society of Mechanical Engineers ASME from [23].

2.2.3. High-Fidelity Model — FE Model

The FE LSP-process model, set up to calculate residual stresses in AA2198 [29] and adopted to AA2024 [37] in the author's previous works, is used in this work to generate a database with the plasma pressure as input and residual stress profiles as output, see Figure 3. The LSP process model consists of a cuboid with dimensions of 60 mm × 60 mm × 4.8 mm and the depth is discretized with an element size of 0.02 mm next to the surface. Sides parallel to *x-z* and *y-z* plane are modelled with fixed boundary conditions, whereas sides parallel to *x-y* plane are considered as free surfaces. The plasma pressure caused by a single laser pulse is modelled as a time dependent surface traction that is uniformly distributed within the peened area. The temporal pressure profile is varied to set up the data set for training, validation and testing. A square of 3 × 3 laser pulses is simulated without overlap, where the square focus size is 3 mm × 3 mm. Residual stresses below the centred laser pulse are averaged layer-wise to calculate a residual-stress-over-depth profile, which has shown to be valid by comparison to experiments [29,37]. The LSP process model consists of approximately 1.4×10^6 continuum elements with reduced integration (C3D8R). The Johnson–Cook material model [38] is utilized, where the used material parameters for AA2024 are summarized in Table 1 for convenience. Nine pressure pulses are simulated in Abaqus/Explicit. A relaxation time of 50 µs is simulated between each pulse, which ensures that the dynamic process reaches a state sufficiently close to equilibrium to prevent significant interaction between two consecutive laser pulses, modelled as pressure loadings. After the simulation of all laser pulses, a final quasi-static implicit simulation (Abaqus/Standard) is conducted to determine the residual stress equilibrium. For further details on the model, the interested reader is referred to [29].

Table 1. Elastic and Johnson–Cook material parameter representative for aluminium alloy AA2024 in T3 heat treatment condition with an equivalent plastic strain rate $\dot{\varepsilon}_{P,0} = 2 \times 10^{-4}\,\mathrm{s}^{-1}$ according to [39].

Parameter	Symbol	Unit	Value
Density	ρ	g/cm^3	2.8
Young's modulus	E	GPa	74
Poisson's ratio	ν	–	0.33
Quasi-static yield strength	A	MPa	350
Strengthening coefficient	B	MPa	972
Strain hardening exponent	n	–	0.73
Dynamic strain hardening coefficient	C	–	0.01

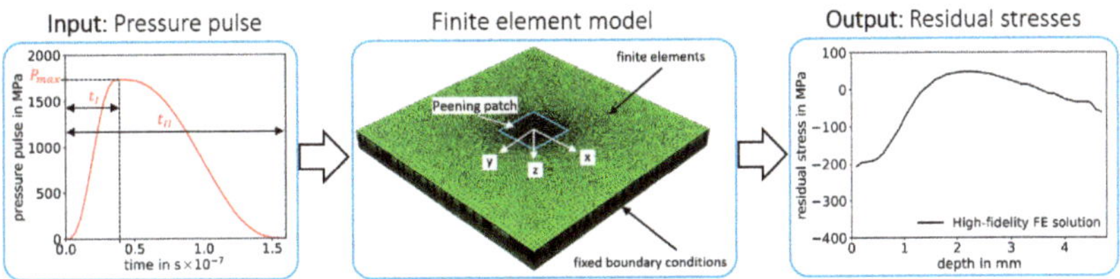

Figure 3. Finite element process model for computing residual stresses induced by pressure pulse from Figure 1.

2.3. Artificial Neural Networks

An ANN represents a computational instrument that can "learn" to correctly map an input to an output via the adjustment of weights. The initial idea of the perceptron was to mimic the behaviour of a neuronal cell in the nervous system of the human brain [40]. Feed forward neural networks are multiple perceptrons composing one or more layers of neurons, where each neuron computes an output based on inputs from the previous layer and an inherent non-linear activation function. The signal is processed in an unidirectional forward direction from input to output throughout the network, where the input signal is progressively transformed into an output signal, see Figure 4. ANNs can be trained to approximate any non-linear relationship [41]. Training of such networks is achieved through back propagating error minimization via gradient descent. The error resulting from the difference between current network output and true/desired output (which is known in a supervised learning task) is minimized by adapting the behaviour of individual neurons through adjusting the weights of the connecting edges between those neurons. The learning rate defines the step size per weight update during gradient descent. For the implementation of an adaptive learning rate, different learning rate optimizers are available, such as Adam [42], Momentum [43] or Adagrad [44], among others. Ultimately, the network represents a mapping rule that is based on provided training examples and is only valid for the space contained in those samples; thus, these networks are not suitable for extrapolating predictions outside the training sample domain. A brief description of a feed forward neural network with back propagating error minimization is provided in the following.

Overall, achieving sufficient training and validation of an ANN depends on the amount of available data, network complexity and the nonlinear nature of the particular relationships to be approximated. To obtain a good ability of the ANN to generalize well, the prediction error on training and validation data sets need to be both low and similar [45], as it indicates that neither underfitting nor overfitting has occurred during training. To prevent overfitting on the training data, learning can be terminated based on the "early stopping" criterion, which is fulfilled as soon as the prediction performance on the validation data set (outside the training data set) is no longer improved during training, even though the error on the training set is still decreasing.

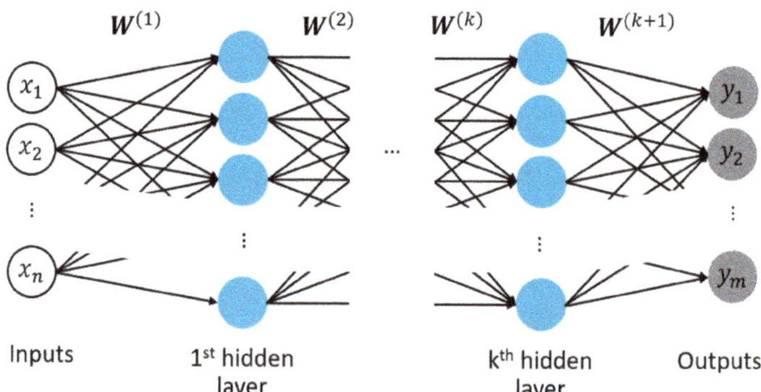

Figure 4. Schematic of a multi-layered neural network with input layer, k hidden layer and output layer, including weight vectors W of edge connections between neurons of adjacent layers for correlating n number of inputs $[x_1, x_2, ..., x_n]$ to m number of outputs $[y_1, y_2, ..., y_m]$.

3. Methodology

First, patterns are generated with pressure pulses and residual stresses from both semi-analytical and FE models. The resulting pairs of semi-analytically and numerically determined residual stress profiles compose the training data set for the corrective task of the ANN. Second, the ANN is trained, validated and tested. Third, the ANN is utilized for correcting semi-analytical residual stress profiles generated by an expanded pulse parameter range that was not contained in the previously utilized training, validation and test data sets. This methodology is described in detail in the following.

As illustrated in Figure 5, the corrected predictions for LSP-induced residual stresses contain the estimates from the physics-based semi-analytical model and a corrective term from the corrective ANN that accounts for the deviation between semi-analytical stresses and numerical stresses to generate the desired high-fidelity solution.

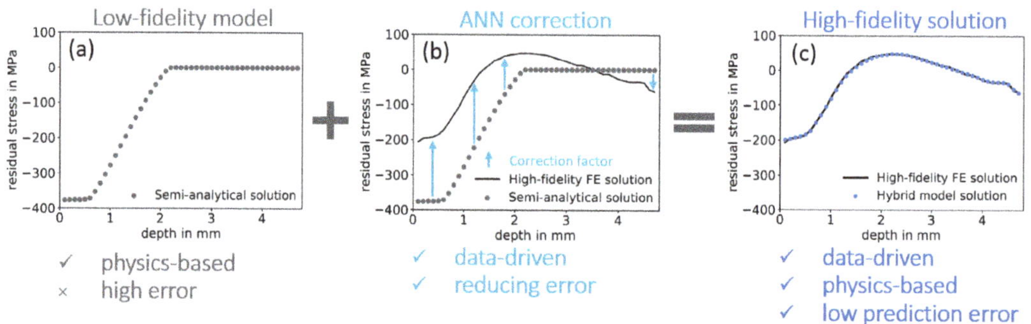

Figure 5. Schematic of hybrid model implementation for prediction of laser shock peening (LSP)-induced residual stresses: (**a**) Residual stresses predicted by the semi-analytical model exhibiting relatively high prediction errors compared to the high fidelity FE solution which is compensated by (**b**) a correction factor "learned" by an artificial neural network (ANN), leading to (**c**) the validated high-fidelity prediction with low errors, i.e., the hybrid model solution.

3.1. Data Preparation

For correcting the coherent residual stress profiles, stress values are discretized over the depth in the form $d(i)/d_{max}$, leading to 47 points from $d(0.1)$ to $d(4.7)$, as surface stresses are disregarded (specimen thickness = 4.8 mm). Included in the input is the information of the known pressure pulses used to generate each residual stress profile with the semi-analytical model. The maximum pressure P_{max} of the particular pulse serves as normalization for all stress values of the respective profile. Since residual stress profiles can converge towards zero, a division by zero or very small stress values during normalization is prevented by a uniform shift of all residual stress values above zero by adding twice the material's yield strength (note that the quasi-static yield strength A is used, here) denoted with σ_y, see Equations (1) and (2). To enable the prediction of correction factors that produce results of high accuracy, one point of the depth discretization is considered at a time. Thus, the depth at which the correction factor for the residual stresses shall be determined is used as the final input. This yields the following dimensionless input space, including shifted and scaled residual stresses over depth:

$$X^i := \left\{ \frac{\sigma^i_{ana,1} + 2\sigma_y}{P^i_{max}}, \frac{\sigma^i_{ana,2} + 2\sigma_y}{P^i_{max}}, ..., \frac{\sigma^i_{ana,47} + 2\sigma_y}{P^i_{max}}, \frac{j}{47} \right\} \tag{1}$$

with i as the sample number, j as the discretization step of the depth in the range from 0.1 mm to 4.7 mm and P^i_{max} as the maximum pressure of the specific sample. The dimensionless output is the correction factor and defined as:

$$Y^i := \left\{ \frac{\sigma^i_{ana,j} + 2\sigma_y}{\sigma^i_{FE,j} + 2\sigma_y} \right\} \tag{2}$$

with $\sigma^i_{ana,j}$ and $\sigma^i_{FE,j}$ being the residual stresses at the depth $j/47$, computed by semi-analytical model and finite element model, respectively. Using a single output, where each output corresponds to a different depth $j/47$, one can observe a smooth curve as a result for the complete continuous distribution since the ANN is forced to smoothly approximate this depth dependency. A smooth curve of the output is obtained when the input is scanned with $j/47$ through the depth. With each j value, a corresponding stress correction at the output is received. The use of physically normed inputs and outputs allows for making predictions in a much wider process parameter range than that used for training of the ANN [46], which can be highly beneficial.

A total number of 82 numerical and semi-analytical sample pairs with pressure pulse parameter ranges listed in Table 2, have been utilized. With the proposed depth discretization of 47, this led to a total of $82 \times 47 = 3854$ patterns that composed the complete data set. The data is randomly split into training, validation and test data sets with an 80/10/10 ratio with the constraint of a stratified P_{max} value range into eight classes, i.e., equidistant subintervals from 800 MPa to 2200 MPa. Thereby, each class is represented in the respective data sets to ensure the ranges of maximum pulse values are similar in training, validation and test data sets, respectively. Ultimately, training, validation and test data sets consisted of 3102, 376 and 376 patterns, respectively. Scaling of inputs and outputs was executed to remain in value ranges of $[-1, 1]$ and $[1, 5]$, respectively. The corrected residual stresses are obtained by solving Equation (2) with respect to absolute values $\sigma^i_{ana,j}$.

Table 2. Pressure pulse parameter ranges of maximum pressure P_{max}, time of maximum pressure t_I and pulse duration t_{II} for training, validation and test data sets.

	P_{max} [MPa]	t_I [ns]	t_{II} [ns]
Min.	800	12	43
Max.	2200	66	300

3.2. Hyperparameters of ANN

The ANN consists of two hidden layers each containing 30 neurons, respectively. The sigmoid function is utilized as the activation function of each layer, except for the final layer, where a linear activation function is implemented to obtain continuous values in the proposed regression task. Gradient descent during mean squared error (MSE)-loss optimization through weight adjustments is enhanced with an adaptive learning rate according to the Adam optimizer. Furthermore, early stopping is implemented to enable training without any overfitting, as training is stopped as soon as the generalization error, i.e., MSE-loss on validation data set, is not decreased any further. Before early stopping is executed, a patience of 1000 further epochs is used to assure that no local minimum on the validation set MSE within this consecutive 1000-epoch-range leads to the stopping. The workflow of this study, consisting of data pre-processing, ANN development and result analysis, has been executed with the open-source libraries Scikit-learn and Keras in conjunction with JupyterNotebook frontend and Tensorflow back-end.

4. Development and Evaluation of ANN-Correction Model

The ANN correction model proposed here is developed and evaluated in two steps. First, the input feature space only contains semi-analytical residual stresses distributed over depth, normalized with the maximum of the corresponding pulse pressure, where the correction predictions still exhibit significant errors. Second, the input feature space is enriched with additional salient features according to a consistent dimensionality analysis, which led to a decrease of those prediction errors. The prediction performances are evaluated with two metrics: Determination coefficient (R^2) and mean squared error (MSE). R^2 is defined as

$$R^2 = 1 - \frac{\sum_{i=1}^{N}\left(y_i - y_{i,pred}\right)^2}{\sum_{i=1}^{N}\left(y_i - y_{mean}\right)^2} , \tag{3}$$

where y_i represents the true value, $y_{i,pred}$ the predicted value, y_{mean} the mean of the true values and N the number of sample values. MSE is defined as

$$MSE = \frac{1}{N} \sum_{i=1}^{N}\left(y_i - y_{i,pred}\right)^2. \tag{4}$$

4.1. Approach 1: Consideration of Only Semi-Analytical Residual Stresses as Input

In this first approach, the input for the corrective ANN prediction consists only of the semi-analytically determined residual stresses, normalized with the maximum pressure value of the pulse, Equation (1). The so-called "learning curves", i.e., values of the loss function (the MSE) on training and validation data sets during training (over the number of epochs), shown in Figure 6a, indicate a significantly lower MSE for predictions on the training data than on the validation data. In other words, the network has been over-fitted to the training data and shows low ability to generalize well, as the prediction error is increased on data points outside the training data set. Correspondingly, the R^2 values for the correction factor, presented in Figure 6b, and the resulting residual stresses, shown in Figure 6c, exhibit deviations between true/desired values and predicted values. Specifically, R^2 values for the correction factor, Equation (2), reached 97.08%, 96.65% and 94.94% on training, validation and test set, respectively, see Table 3. For the predictions of corrected residual stresses, these deviations are even greater, with R^2 values of 91.14%, 91.35% and 81.88% for training, validation, and test sets, respectively, see Table 3. Comparisons of input, output and corrected residual stresses of three exemplary test samples are shown in Figure 7, where the corrections of the semi-analytical stresses are not in good agreement with the desired FE solutions. The error of the stress predictions is decreased through the correction but not to a satisfactory extend. In order to improve corrective model predictions with respect to an increased determination coefficient R^2 and a decreased MSE, additional information needs to be provided in the input space for the ANN.

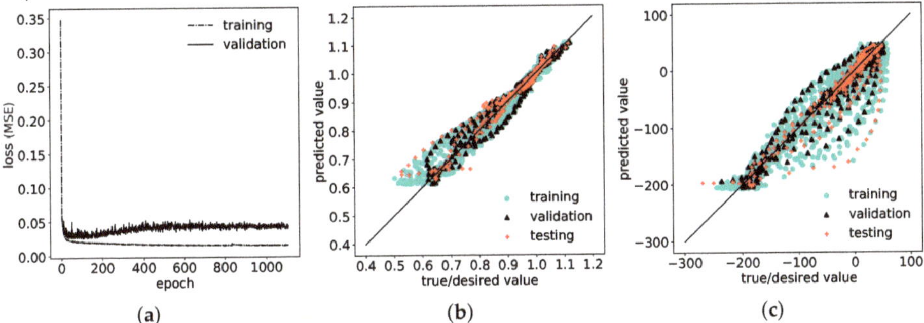

Figure 6. (**a**) Learning curves: Mean squared error (MSE)-loss function values minimized via weight adjustment of the ANN on training set and simultaneous MSE for predictions on validation set with training-set weights over number of epochs during training. (**b**) Determination coefficient R^2 for correction factor (ANN output) achieved by ANN on training, validation and test data sets. (**c**) Determination coefficient R^2 for related residual stresses attained by ANN on training, validation and test data sets.

Table 3. Prediction metrics of trained ANN via Approach 1: R^2 (determination coefficient) and MSE (mean squared error) for correction coefficients as well as for corresponding residual stresses on training, validation and test data sets, respectively.

	Correction Factor		Residual Stresses	
Data Set	R^2 **in %**	*MSE*	R^2 **in %**	*MSE* **in MPa**^{2}
Training	97.08	0.000466	91.14	399.21
Validation	96.65	0.000602	91.35	452.26
Test	94.94	0.000669	81.88	607.42

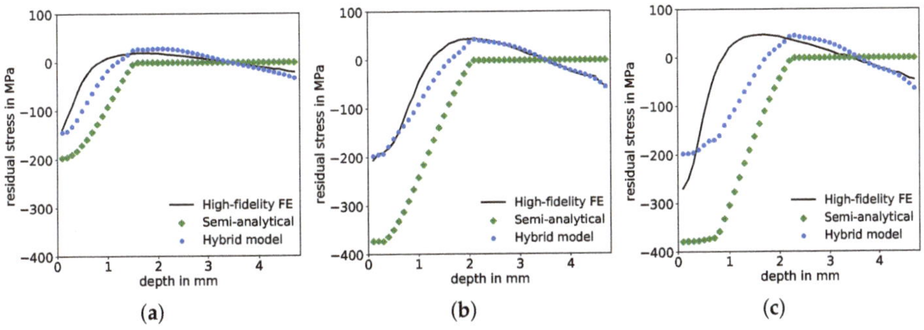

Figure 7. Comparison of residual stress distributions over depth predicted by the FE model, semi-analytical model and hybrid model for three exemplary test samples with pulse parameters maximum pressure P_{max}, time of maximum pressure t_I and pulse duration t_{II} of (**a**) 1236 MPa, 15.1 ns, 85 ns; (**b**) 1639 MPa, 37.7 ns, 145 ns; and (**c**) 1820 MPa, 13 ns, 65.7 ns.

As mentioned in Section 2.2.1, the pulse duration t_{II} is not considered in the semi-analytical model according to its input definition, the pressure pulse duration is only considered until t_I. As a result, samples whose corresponding pressure pulses differ uniquely only in duration will cause predictions of identical residual stress distributions, see Figure 8. Mathematically, this is a non-injective relationship, inadequate to be represented by any function, i.e., the same input could certainly not be correlated to multiple different outputs via the ANN-model of this first approach, where only residual stresses over depth serve as input. Consequently, as pulse duration t_{II} affects the prediction result, it needs to be considered in the input space for the corrective model.

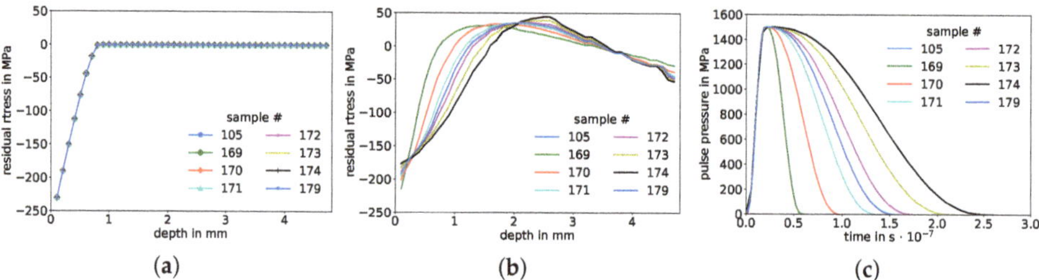

Figure 8. (**a**) Super-imposed but indistinguishable residual stress distributions over depth predicted by the semi-analytical model for different pressure pulses, i.e., identical inputs for the corrective ANN-model. (**b**) Corresponding output targets: Eight unique residual stress distributions over depth predicted by the FE model and (**c**) corresponding distinctive pressure pulses over time that were used as input for both models, exhibiting different pulse durations but identical maximum pressures and times of respective maximum pressures.

4.2. Approach 2: Adding Salient Features to the Input Space

In order to enable a unique mapping between inputs and outputs, additional input features are identified via a dimensionality analysis and are added to the input space. In accordance with the Buckingham Π theorem [47], a required minimum number of dimensionless parameters can be defined to sufficiently describe the physical problem. Thus, besides the analytical stresses σ_{ana} and maximum pressure P_{max}, the pressure pulse time quantities t_I, t_{II} and t_{III} are included. To connect those temporal measures to mechanical properties E and ρ, the wave speed $c = \sqrt{E/\rho}$ is also considered. Ultimately, the peened area A_{peened} is used to complete the set of five dimensionless quantities:

$$\Pi_1 = \frac{\sigma_{ana}}{P_{max}}, \ \Pi_2 = \frac{t_I}{t_{II}}, \ \Pi_3 = \frac{t_{III}}{t_{II}}, \ \Pi_4 = t_{III}\sqrt{\frac{E}{\rho \cdot A_{peen}}}, \ \Pi_5 = \frac{P_{max}}{E}. \tag{5}$$

Adding dimensionless information that is based on a consistent dimensionality analysis to the input space leads to a reduction of inaccuracies, which is in agreement with a study based on a similar input definition for an ANN [46]. Subsequently, this leads to a further reduction of prediction's MSE and increase of R^2 compared to the first approach presented in Section 4.1. All input-output pairs can be uniquely identified by the ANN. Accordingly, the modified input is described with:

$$X^i := \left\{ \frac{\sigma_{ana,1}^i + 2\sigma_y}{P_{max}^i}, \frac{\sigma_{ana,2}^i + 2\sigma_y}{P_{max}^i}, ..., \frac{\sigma_{ana,47}^i + 2\sigma_y}{P_{max}^i}, \frac{t_I^i}{t_{II}^i}, \frac{t_{III}^i}{t_{II}^i}, t_{III}^i\sqrt{\frac{E}{\rho \cdot A_{peen}}}, \frac{P_{max}^i}{E}, \frac{j}{47} \right\}. \tag{6}$$

This dimensionless formulation ensures that all dependencies are scaled without loss of generality. In comparison to the first approach, the previous bias and variance indicated in the learning curves in Figure 6a is reduced, as the final MSE-loss is further reduced on both training as well as on validation data sets, respectively, and both converged towards similar values, see Figure 9a. Hence, prediction results improved significantly on all three data sets, with respect to increased determination coefficients R^2 each above 99% for the correction factors, see Figure 9b and also for the corrected residual stresses, see Figure 9c. The MSE on the test set declined simultaneously to a maximum of 3.9×10^{-5} and 28.63 MPa^2 for correction factors and corrected residual stresses, respectively, see Table 4. There is good agreement between the corrected prediction and the desired values of the residual stresses throughout the complete depth, as demonstrated by three examples from the test data set in Figure 10.

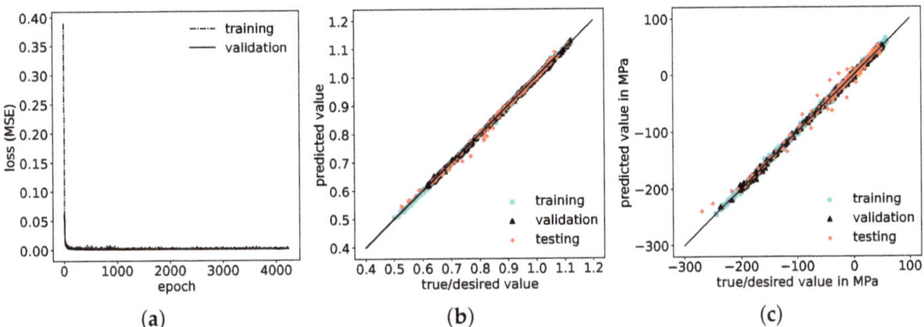

Figure 9. (**a**) Learning curves: MSE-loss function values on training and validation data sets over number of epochs during training and (**b**) corresponding prediction values of the correction factor versus true values, and of (**c**) the corresponding residual stresses.

Table 4. Prediction metrics of the trained ANN via Approach 2: Determination coefficient R^2 and MSE for correction coefficients as well as corresponding residual stresses achieved on training, validation and test data sets, respectively.

	Correction Factor		**Residual Stresses**	
Data Set	R^2 **in %**	MSE	R^2 **in %**	MSE **in MPa**^{2}
Training	99.95	7×10^{-6}	99.90	4.33
Validation	99.93	12×10^{-6}	99.86	7.38
Test	99.71	39×10^{-6}	99.15	28.63

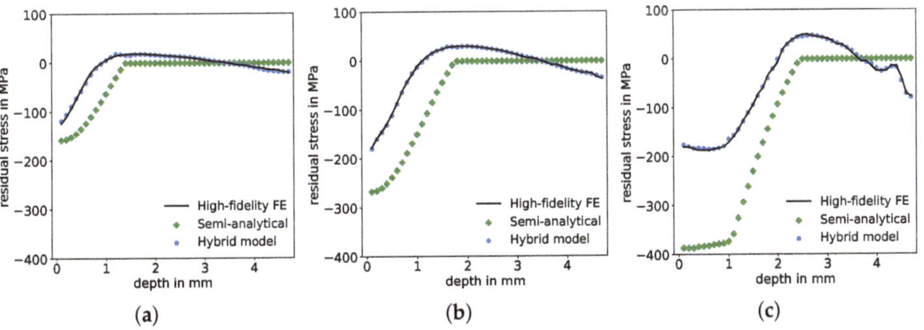

Figure 10. Comparison of residual stress distributions over depth predicted by the FE model, semi-analytical model and hybrid model for three test samples with maximum pressure P_{max}, time of maximum pressure t_I and pulse duration t_{II} of (**a**) 1144 MPa, 38.9 ns, 137 ns; (**b**) 1390 MPa, 22.2 ns, 140 ns; and (**c**) 2039 MPa, 49.5 ns, 243 ns, respectively.

5. Generalization of Hybrid Model

An evaluation of the generalization ability is performed by expanding the input parameter space, i.e. value ranges of pressure pulse parameters: Maximum pressure P_{max}, time of maximum pressure t_I and pulse duration t_{II}, to respective ranges that were not used for training, validation and testing, as shown in Figure 11 and Table 5. The lower bound of the maximum pressure range remained at 800 MPa because there is an almost insignificant contribution to residual stress formation by pressure pulses with a maximum below 800 MPa. In addition, extension of maximum pressures above 2400 MPa becomes physically unfeasible. Ultimately, there is no significant expansion but only minor exceedances for P_{max} values beyond the training space. Lower bounds of pulse durations were decreased from 12 ns to 1 ns and upper bounds increased from 66 ns to 100 ns. The expanded-space data-set contained 35 samples. With this expanded parameter space, deviations between semi-analytical and high-fidelity solutions can be adequately corrected by the ANN and its trained range of correction factors.

The "learned" range for the correction factors is $[0.5090, 1.1189]$; thus, the deviation between analytical and numerical model has to be correctable by values within that range in order to achieve the anticipated solutions. Restrictions are inevitable when the required factor for an appropriate correction lies outside this range. In this case, no correction is performed by the ANN and the analytical input is also the output. This corresponds to setting the correction factor to 1.0. Thus, the default prediction, in a worst-case scenario, is the provided input—the prediction of the semi-analytical model, which can be noticed clearly and used as an indicator for no correction having been performed. Essentially, an extrapolating prediction on an expanded parameter space can only be performed as long as the output of the ANN, i.e., the required correction factor, still lies in the value range of the training data set.

Table 5. Expanded pressure pulse parameter ranges of maximum pressure P_{max}, time of maximum pressure t_I and pulse duration t_{II} as extrapolated parameter space in comparison to the ranges in the data set used for training, validation and testing, see Table 2.

		P_{max} in MPa	t_I in ns	t_{II} in ns
Training, validation, test	Min.	800	12	43
	Max.	2200	66	300
Expanded parameter space	Min.	800	1	43
	Max.	2400	100	306

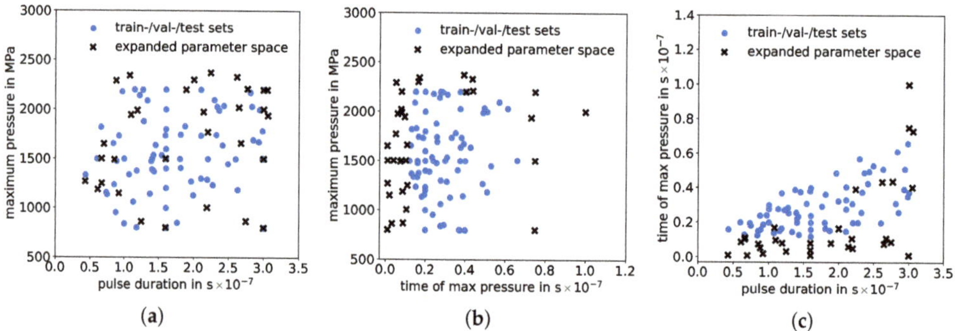

Figure 11. Sample positioning in the expanded parameter space: Maximum pressure over (**a**) pulse duration and over (**b**) time of maximum pressure as well as (**c**) time of maximum pressure over pulse duration.

5.1. Setup of Purely Data-Driven ANN as Benchmark

The prediction performance of the hybrid model is benchmarked against the estimations of a purely data-driven ANN trained directly with pressure-pulse-over-time as input and residual-stresses-over-depth provided by the FE-model as output, without the consideration of any physics-based model. In the following, this purely data-driven ANN is briefly explained. Essentially, no corrective task is performed and the input consists of 47 discretized pressure values and the respective terms defined in the dimensionality analysis with

$$X_{direct}^i := \left\{ \frac{P_1^i}{Pi_{max}}, \frac{P_2^i}{Pi_{max}}, \dots, \frac{P_{47}^i}{Pi_{max}}, \frac{t_I^i}{t_{II}^i}, \frac{t_{III}^i}{t_{II}^i}, t_{III}^i \sqrt{\frac{E}{\rho \cdot A_{peen}}}, \frac{Pi_{max}}{\sigma_y}, \frac{j}{47} \right\}. \tag{7}$$

The output space contains the residual stress values, where constant discretization over specimen depth of the residual stresses is used, similar to the output discretization of the output space for the corrective model by

$$Y^i_{direct} := \left\{ \frac{\sigma^i_{FE,j} + 2\sigma_y}{\sigma_y} \right\} \tag{8}$$

where superscript i refers to the sample number and subscript j to the depth discretization step of 0.1 mm in the range from 0.1 mm to 4.7 mm.

The previous ANN architecture consisting of two hidden layers with respective 30 neuron and sigmoid activation functions is used to avoid any artificial influence of the ANN architecture in the benchmark. Likewise, early stopping is implemented to avoid overfitting during training. Normalization of inputs to $[-1,1]$ and outputs to $[1,5]$ is performed, as for the hybrid model.

5.2. Comparison of Physics-Based Hybrid Model and Purely Data-Driven ANN

With a comparison of the physics-based corrective prediction in Figure 12a to a purely data-driven ANN prediction model in Figure 12b, an example for the benefits of a corrective physics-based ANN model over a purely data-driven ANN is provided. As can be seen in the comparison of the predictions in the expanded parameter space, predictions that are purely based on data exhibit pronounced errors, which is not the case for the ANN where physical laws are considered in the contained analytical model solution. This good prediction performance is a consequence of remaining within the trained range of correction factors as well as a result of the enhanced prediction ability of the hybrid model itself, which is owed to the decreased complexity of the correction problem. Even though the R^2 values for the data-driven ANN are both above 99% on training and validation sets as well as above 95% on the testing set, the MSEs on the expanded space are almost two orders of magnitude higher than the one of the physics-based corrective model and amounts to over 1700 MPa^2, see Table 6. The MSE of the corrective model measures just below 31 MPa^2. The determination coefficient R^2 of the corrective approach on the extrapolation data set is highly alike to the R^2 values on the other data sets and is still as high as 99.39%, whereas for the data-driven approach, the R^2 values are all above 99% on training and validation and above 95% on test data sets but drops down to 65% for predictions on the expanded parameter space.

The absolute value of the relative error of both physics-based corrective model as well as purely data-driven model is defined as *err*, according to [48], and computed via:

$$err := \left| \frac{d - y^N}{d} \right| \tag{9}$$

with true values d, predicted values y and number of samples N. The maximum *err* from the data-driven model is approximately 53% and just below 8% for the corrective model at $n/N = 1$, as shown in Figure 12c, where the normalized number of samples is sorted from small to large *err* values. As a result, consideration of the problem's physics through the semi-analytical model leads to a better generalization compared to using a purely data-driven predictor relying on the relevant physics to be represented (only) in the training data. In particular, via the corrective ANN, interpolation within its trained value range of correction factors can still be performed, even on the expanded parameter space; whereas via the data-driven ANN, extrapolating predictions are performed within the expanded parameter space, which is unfeasible for an ANN because its predictive function is fitted to the training data and becomes unreliable in a variable space for which no training data is available. So, in this use-case example, based on a physically reasonable extension of the parameter space, a physics-based correction model exhibits superior prediction performance over a data-driven model, under the condition that results can be adjusted with the trained range of correction factors to achieve the anticipated solution.

Table 6. Prediction metrics of the hybrid model and purely data-driven ANN: R^2 and MSE for residual stresses of samples in training, validation, test and expanded parameter space data sets.

Data Set	Hybrid Model		Data-Driven ANN	
	R^2 in %	MSE	R^2 in %	MSE in MPa^2
Training	99.90	4.33	99.86	6.32
Validation	99.86	7.38	99.76	12.39
Test	99.15	28.63	95.89	137.58
Expanded space	99.39	30.17	65.00	1717.18

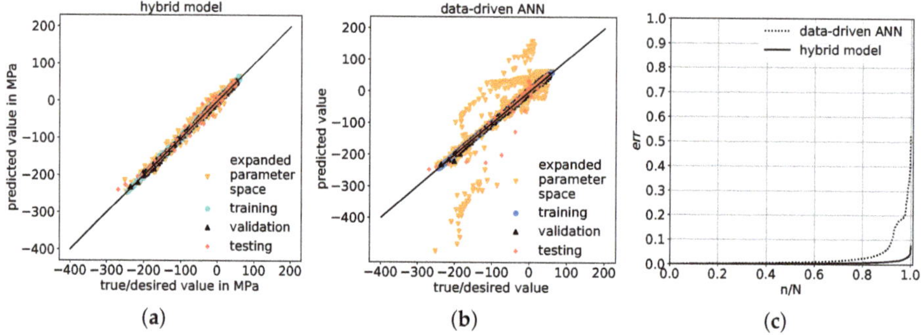

Figure 12. Juxtaposition of predicted values and true/desired values on training, validation, test sets and expanded parameter space data set, achieved by (**a**) the physics-based hybrid model and (**b**) the purely data-driven ANN, respectively. (**c**) shows the relative error of samples n normalized with the total number of samples N, sorted from low to high *err* values on the data set with expanded parameter space generated by hybrid model and data-driven ANN.

5.3. Data Reduction Effects on Hybrid Model and Data-Driven ANN Predictions

In this section, the prediction performances of the hybrid model and the data-driven ANN are juxtaposed while the total number of samples is reduced. The total data set is split into training, validation and test data sets via a constant data-split ratio of 80/10/10, throughout a reduction of the total data set from 100% to 20% by increments of 10%. Thus, a 100% data set consists of 66 training, 8 validation and 8 test samples (as in all previous sections); whereas a 20% data set contains 13 training, 1 validation and 1 test sample(s). The specific samples and total sample number in the expanded-space data set remained constant at 35. For each data-reduction step, the data split is performed randomly and three times, each time with a different random state, in order to avoid prediction results that depend on specific samples contained in the respective data sets. Consequently, the MSE average and standard deviation of the corresponding three prediction models are calculated and used for further evaluation.

The hybrid model outperforms the data-driven ANN on the test data set with respect to an overall decreased mean MSE and continuously lower standard deviations. On average, the mean MSE is lower and its standard deviation decreased, when performing predictions with the hybrid model compared to the data-driven ANN. As shown in Figure 13a, these outperformances appear clearly once the amount of samples in the total data set in reduced below 60%, i.e., below a sample number of 49, as well as at the smallest total data set of 20%, respectively. On the extrapolation data set, the superior prediction ability of the hybrid model over the data-driven ANN is magnified with respect to a significantly lower average mean MSE and a substantially decreased standard deviation, see Figure 13b.

These outperformances could be due to several reasons. Primarily, the corrective ANN with its correction factor prediction is assumed to be more simple in complexity and in non-linearity than the residual stress prediction of the data-driven ANN. Consequently, the corrective ANN in combination with the semi-analytical model is more stable and robust in its predictions once the amount of data is reduced, in comparison to the data-driven ANN. In addition, there appears to be a higher dependence on specific samples being contained in the training and validation data sets for the data-driven ANN since the variation of mean MSE and standard deviation are more significant within an identical amount of data (but different random data splits). Ultimately, the proposed corrective approach, i.e., hybrid model consisting of the semi-analytical model and the corrective ANN, exhibits a number of benefits over a purely data-driven ANN, even more when the amount of data is scare or very limited, such as in DoE data sets.

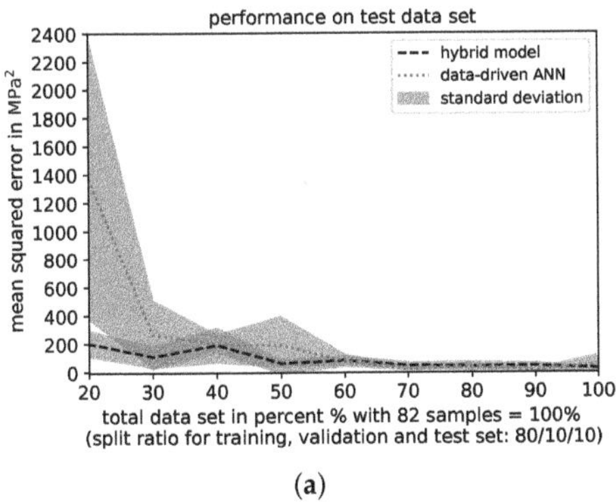

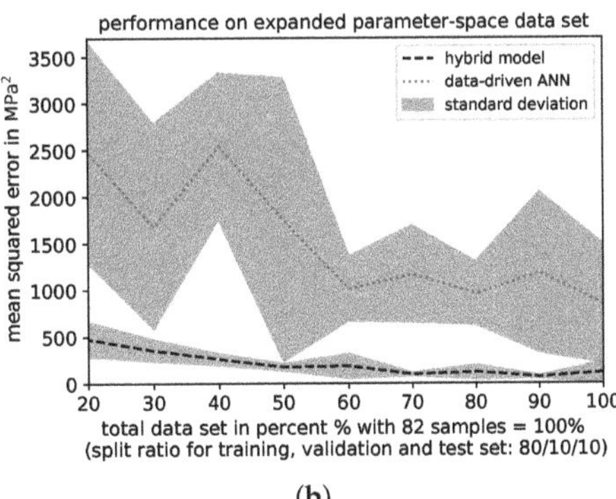

Figure 13. Comparison of prediction performances of hybrid model and direct ANN with respect to the average mean squared error (MSE) and standard deviation achieved on (**a**) the test data set and (**b**) the extrapolation data set, while reducing the amount of the total data set (training, validation and test data sets) from 100% to 20% in increments of 10%-steps, respectively. All MSE average values and standard deviations are based on three different MSEs and their respective standard deviations that are achieved on dissimilar data splits implemented by changing pseudo-random-states.

6. Conclusions

In this study, a physics-based semi-analytical model, representing a rather simple but very efficient model, has been successfully combined with a corrective ANN into an hybrid prediction model to enhance its applicability range. Ultimately, low prediction errors were reached with respect to the desired high-fidelity solution, provided by a numerical FE simulation in the investigated use case of LSP. The high-fidelity numerical data could easily be replaced by experimental data, enabling correction towards empirical measurements. A number of prerequisites for adequately performing the correction task have been identified. Primarily, unique relationships between inputs and outputs need to exist, where redundancies in the data can be an indicator for non-unique relationships. These non-unique relationships may be compensated by using additional (salient) features identified via a consistent dimensionality analysis. Upon detectable uniqueness between inputs and outputs, low prediction errors are enabled. Essential findings for achieving a low prediction error in our specific problem domain are:

- Through the proposed corrective approach of a semi-analytical model, the solution of a high-fidelity numerical simulation is reached very efficiently.
- In particular, trained range of correction factors allows for a maximum adjustments of semi-analytical stresses of up to approximately 50% towards the desired high-fidelity solution.
- Generalized predictions for extended process parameter ranges can be achieved under the condition of correction factor values remaining within the training value range.

- Within the value range of trained correction factors, the generalization of the physics-based corrective approach within an expanded-parameter-space performs with significantly lower prediction errors compared to a purely data-driven generalization.
- When reducing the amount of available data during training, validation and testing, the generalization via the corrective approach demonstrated significantly reduced prediction errors compared to the purely data-driven model on both test set and expanded parameter-space data set, illustrating its ability to handle sparse data.

Author Contributions: Conceptualization, F.E.B., N.H. and B.K.; methodology, F.E.B., S.K., N.H. and B.K; validation, F.E.B. and S.K.; resources, N.H. and B.K.; data curation, F.E.B. and S.K.; formal analysis, F.E.B., S.K. and N.H.; software, F.E.B. and S.K.; writing—original draft preparation, F.E.B.; writing—review and editing, F.E.B., S.K., N.H., B.K.; visualization, F.E.B. and S.K.; supervision, N.H. and B.K.; project administration, B.K. All authors have read and agreed to the published version of the manuscript.

Funding: This research received no external funding.

Data Availability Statement: The data presented in this study are available upon reasonable request from the corresponding author.

Acknowledgments: The authors thank Yongxiang Hu from the Department of Mechanical Engineering of the Shanghai Jiao Tong University for providing the code of the semi-analytical model, published in [23].

Conflicts of Interest: The authors declare no conflict of interest.

References

1. Kalidindi, S.R.; Graef, M. Materials Data Science: Current Status and Future Outlook. *Annu. Rev. Mater. Res.* **2015**, *45*, 171–193.
2. Bock, F; Aydin, R.; Cyron, C.; Huber, N.; Kalidindi, S.R., Klusemann, B. A Review of the Application of Machine Learning and Data Mining Approaches in Continuum Materials Mechanics. *Front. Mater.* **2019**, *6*, 443.
3. Altschuh, P.; Yabansu, Y.C.; Hötzer, J.; Selzer, M.; Nestler, B.; Kalidindi, S.R. Data science approaches for microstructure quantification and feature identification in porous membranes. *J. Membr. Sci.* **2017**, *540*, 88–97.
4. Brough, D.B.; Kannan, A.; Haaland, B.; Bucknall, D.G.; Kalidindi, S.R. Extraction of Process-Structure Evolution Linkages from X-ray Scattering Measurements Using Dimensionality Reduction and Time Series Analysis. *Integr. Mater. Manuf. Innov.* **2017**, *6*, 147–159.
5. Yun, M.; Argerich, C.; Cueto, E.; Duval, J.L.; Chinesta, F. Nonlinear Regression Operating on Microstructures Described from Topological Data Analysis for the Real-Time Prediction of Effective Properties. *Materials* **2020**, *13*, 2335.
6. Adams B.L.; Kalidindi, S.R.; Fullwood, D.T. *Microstructure Sensitive Design for Performance Optimization*; Elsevier: Amsterdam, The Netherlands, 2013.
7. Cecen, A.; Dai, H.; Yabansu, Y.C.; Kalidindi, S.R.; Le, S. Material structure-property linkages using three-dimensional convolutional neural networks. *Acta Mater.* **2018**, *146*, 76–84.
8. Chupakhin, S.; Kashaev, N.; Klusemann, B.; Huber, N. Artificial neural network for correction of effects of plasticity in equibiaxial residual stress profiles measured by hole drilling. *J. Srain. Anal. Eng.* **2017**, *52*, 137–151.
9. Bock, F.E.; Blaga, L.A.; Klusemann, B. Mechanical Performance Prediction for Friction Riveting Joints of Dissimilar Materials via Machine Learning. *Procedia Manuf.* **2020**, *47*, 615–622.
10. Yang, Z.; Yabansu, Y.C.; Al-Bahrani, R.; Liao, W.K.; Choudhary, A.N.; Kalidindi, S.R.; Agrawal, A. Deep learning approaches for mining structure-property linkages in high contrast composites from simulation datasets. *Comput. Mater. Sci.* **2018**, *151*, 278–287.
11. Raissi, M. Deep Hidden Physics Models: Deep Learning of Nonlinear Partial Differential Equations. *J. Mach. Learn. Res.* **2018**, *18*, 1–24.
12. Lu, L.; Dao, M.; Kumar, P.; Ramamurty, U.; Karniadakis, G.E.; Suresh, S. Extraction of mechanical properties of materials through deep learning from instrumented indentation. *Proc. Natl. Acad. Sci. USA* **2020**, *117*, 7052–7062.
13. Chinesta, F; Cueto, E.; Abisset-Chavanne, E.; Duval, J.L.; Khaldi, F.E.B. Virtual, Digital and Hybrid Twins: A New Paradigm in Data-Based Engineering and Engineered Data. *Arch. Comput. Methods Eng.* **2018**, *27*, 105–134.
14. Montáns, F.J.; Chinesta, F.; Gómez-Bombarelli, R.; Kutz, J.N. Data-driven modeling and learning in science and engineering *Comptes Rendus MéCanique* **2019**, *347*, 845–855.
15. Kirchdoerfer, T; Ortiz, M. Data-driven computational mechanics. *Comput. Methods Appl. Mech. Eng.* **2016**, *304*, 81–101.
16. Karpatne, A.; Atluri, G.; Faghmous, J.H.; Steinbach, M.; Banerjee, A.; Ganguly, A.; Shekhar, S.; Samatova, N.; Kumar, V. Theory-Guided Data Science: A New Paradigm for Scientific Discovery from Data *IEEE Trans. Knowl. Data Eng.* **2017**, *29*, 2318–2331.

17. Liu, X.; Athanasiou, C.E.; Padture, N.P.; Sheldon, B.W.; Gao, H. A machine learning approach to fracture mechanics problems. *Acta Mater.* **2020**, *190*, 105–112.
18. Kapteyn, M.G.; Knezevic, D.J.; Huynh, D.B.P.; Tran, M.; Willcox, K.E. Data-driven physics-based digital twins via a library of component-based reduced-order models *Int. J. Numer. Methods Eng.* **2020**, *53*, 3073.
19. Moya, B.; Badías, A.; Alfaro, I.; Chinesta, F.; Cueto, E. Digital twins that learn and correct themselves *Int. J. Numer. Methods Eng.* **2020**, *25*, 87.
20. González, D; Chinesta, F; Cueto, E. Learning Corrections for Hyperelastic Models From Data. *Front. Mater.* **2019**, *6*, 752.
21. Ibáñez, R.; Abisset-Chavanne, E.; González, D; Duval, J.L.; Cueto, E; Chinesta, F. Hybrid constitutive modeling: Data-driven learning of corrections to plasticity models. *Int. J. Mater. Form.* **2019**, *12*, 717–725.
22. Havinga, J.; Mandal, P.K.; van den Boogaard, T. Exploiting data in smart factories: Real-time state estimation and model improvement in metal forming mass production. *Int. J. Mater. Form.* **2020**, *13*, 663–673.
23. Hu, Y; Yao, Z.; Hu, J. An Analytical Model to Predict Residual Stress Field Induced by Laser Shock Peening. *J. Manuf. Sci. Eng.* **2009**, *131*, 031017.
24. Dursun, T.; Soutis, C. Recent developments in advanced aircraft aluminium alloys. *Mater. Des.* **2014**, *56*, 862–871.
25. Hertwich, E.G.; Ali, S.; Ciacci, L.; Fishman, T.; Heeren, N.; Masanet, E.; Asghari, F.N.; Olivetti, E.; Pauliuk, S.; Tu, Q.; Wolfram, P. Material efficiency strategies to reducing greenhouse gas emissions associated with buildings, vehicles, and electronics—A review. *Environ. Res. Lett.* **2019**, *14*, 043004.
26. Peyre, P.; Fabbro, R. Laser shock processing: A review of the physics and applications. *J. Mater. Process. Technol.* **1995**, *27*, 1213–1229.
27. Braisted, W.,Brockman, R. Finite element simulation of laser shock peening. *Int. J. Fatigue* **1999**, *21*, 719–724.
28. Brockman, R.A.; Braisted, W.R.; Olson, S.E.; Tenagli, R.D.; Clauer, A.H.; Langer, K.; Shepard, M.J. Prediction and characterization of residual stresses from laser shock peening. *Int. J. Fatigue* **2012**, *36*, 96–108.
29. Keller, S.; Chupakhin, S.; Staron, P.; Maawad, E.; Kashaev, N.; Klusemann, B. Experimental and numerical investigation of residual stresses in laser shock peened AA2198. *J. Mater. Process. Technol.* **2018**, *255*, 294–307.
30. Frija, M; Ayeb, M.; Seddik, R.; Fathallah, R.; Sidhom, H. Optimization of peened-surface laser shock conditions by method of finite element and technique of design of experiments. *Int. J. Adv. Manuf. Technol.* **2018**, *97*, 51–69.
31. Ayeb, M.; Frija, M; Fathallah, R. Prediction of residual stress profile and optimization of surface conditions induced by laser shock peening process using artificial neural networks. *Int. J. Adv. Manuf. Technol.* **2019**, *100*, 2455–2471.
32. Wu, J.; Li, Y.; Zhao, J.; Qiao, H.; Lu, Y.; Sun, B; Hu, X.; Yang, Y. Prediction of residual stress induced by laser shock processing based on artificial neural networks for FGH4095 superalloy. *Mater. Lett.* **2021**, *286*, 129269.
33. Mathew, J.; Kshirsagar, R.; Zabeen, S.; Smyth, N.; Kanarachos, S.; Langer, K.; Fitzpatrick, M.E. Machine Learning-Based Prediction and Optimisation System for Laser Shock Peening. *Appl. Sci.* **2021**, *11*, 2888.
34. Ebert, S.D.; Kenton Musgave, F.; Peachey, D.; Perlin, K.; Worley, S. *Texturing & Modeling—A Procedural Approach*, 3rd ed.; Morgan Kaufmann Series in Computer Graphics and Geometric Modeling: San Francisco, CA, USA, 2003; pp. 30–31.
35. Timoshenko, S.; Goodier, J. *Theory of Elasticity*, 2nd ed.; McGraw-Hill: New York, NY, USA, 1951.
36. Mcdowell, D. L. An Approximate Algorithm for Elastic-Plastic Two-Dimensional Rolling/Sliding Contact. *Wear* **1997**, *211*, 237–246.
37. Keller, S.; Horstmann, M; Kashaev, N.; Klusemann, B. Experimentally validated multi-step simulation strategy to predict the fatigue crack propagation rate in residual stress fields after laser shock peening. *Int. J. Fatigue* **2019**, *124*, 265–276.
38. Johnson, G.R.; Cook, W.H. A constitutive model and data for metals subjected to large strains, high strain rates and high temperatures. In Proceedings of the 7th International Symposium on Ballistics, The Hague, The Netherlands, 19–21 April 1983; Volume 21, pp. 541–547.
39. Sticchi, M.; Staron, P.; Sano, Y.; Meixer, M.; Klaus, M.; Rebelo-Kornmeier, J.; Huber, N.; Kashaev, N. A parametric study of laser spot size and coverage on the laser shock peening induced residual stress in thin aluminium samples. *J. Eng.* **2015**, *13*, 97–105.
40. Rosenblatt, F. The perceptron: A probabilistic model for information storage and organization in the brain. *Psychol. Rev.* **1958**, *65*, 386–408.
41. Haykin, S. *Neural Networks. A Comprehensive Foundation*; 2nd ed.; Prentice Hall: Upper Saddle River, NJ, USA, 1998.
42. Kingma, D.; Ba, J. Adam: A Method for Stochastic Optimization. *arXiv* **2014**, arXiv:1412.6980.
43. Qian, N. On the momentum term in gradient descent learning algorithms. *Neural Netw.* **1999**, *12*, 145–151.
44. Duchi, J.; Hazan, E.; Singer, Y. Adaptive Subgradient Methods for Online Learning and Stochastic Optimization. *J. Mach. Learn. Res.* **2011**, *12*, 2121–2159.
45. Mitchell, T. *Machine Learning*, 2nd ed.; McGraw-Hill: New York, NY, USA, 2010; p. 67.
46. Huber, N.; Tsakmakis, C. A new loading history for identification of viscoplastic properties by spherical indentation. *J. Mater. Res.* **2004**, *19*, 101–113.
47. Gibbings, J.C. *Dimensional Analysis*; Springer: London, UK; New York, NY, USA, 2011.
48. Huber, N.; Tsakmakis, C. Determination of constitutive properties from spherical indentation data using neural networks. Part I: The case of pure kinematic hardening in plasticity laws. *J. Mech. Phys. Solids* **1999**, *47*, 1569–1588.

MDPI
St. Alban-Anlage 66
4052 Basel
Switzerland
Tel. +41 61 683 77 34
Fax +41 61 302 89 18
www.mdpi.com

Materials Editorial Office
E-mail: materials@mdpi.com
www.mdpi.com/journal/materials